T0231360

POLLUTION AND THE ATMOSPHERE
Designs for Reduced Emissions

POLLUTION AND THE ATMOSPHERE

Designs for Reduced Emissions

Edited by
Marco Ragazzi, PhD

APPLE ACADEMIC PRESS

Apple Academic Press Inc. | Apple Academic Press Inc.
3333 Mistwell Crescent | 9 Spinnaker Way
Oakville, ON L6L 0A2 | Waretown, NJ 08758
Canada | USA

©2017 by Apple Academic Press, Inc.

First issued in paperback 2021

Exclusive worldwide distribution by CRC Press, a member of Taylor & Francis Group
No claim to original U.S. Government works

ISBN 13: 978-1-77-463678-7 (pbk)
ISBN 13: 978-1-77-188513-3 (hbk)

Library and Archives Canada Cataloguing in Publication

Pollution and the atmosphere : designs for reduced emissions/edited by Marco Ragazzi, PhD.

Includes bibliographical references and index.
Issued in print and electronic formats.
ISBN 978-1-77188-513-3 (hardcover).--ISBN 978-1-315-36563-3 (PDF)

1. Air--Pollution. 2. Waste gases. 3. Hazardous waste site remediation.
4. Greenhouse gas mitigation. I. Ragazzi, Marco, author, editor

TD885.P64 2016   628.5'32   C2016-907676-8   C2016-907677-6

Library of Congress Cataloging-in-Publication Data

Names: Ragazzi, Marco, editor.
Title: Pollution and the atmosphere : designs for reduced emissions / editor, Marco Ragazzi, PhD.
Description: Toronto ; New Jersey : Apple Academic Press, 2015. | Includes bibliographical references and index.
Identifiers: LCCN 2016052730 (print) | LCCN 2016053985 (ebook) | ISBN 9781771885133 (hardcover : acid-free paper) | ISBN 9781771885140 (eBook) | ISBN 9781315365633 (ebook)
Subjects: LCSH: Air--Pollution. | Waste gases. | Hazardous waste site remediation. | Greenhouse gas mitigation.
Classification: LCC TD885 .P65 2015 (print) | LCC TD885 (ebook) | DDC 628.5/32--dc23
LC record available at https://lccn.loc.gov/2016052730

Apple Academic Press also publishes its books in a variety of electronic formats. Some content that appears in print may not be available in electronic format. For information about Apple Academic Press products, visit our website at **www.appleacademicpress.com** and the CRC Press website at **www.crc-press.com**

About the Editor

MARCO RAGAZZI, PhD

Marco Ragazzi has a PhD in sanitary engineering from Milan Polytechnic, Italy. The author or co-author of more than 500 publications, he is currently associate professor at the Department of Civil, Environmental, and Mechanical Engineering in the University of Trento, Italy. Presently he is in charge for the environmental sustainability policies of the University of Trento. His research interests include solid waste and wastewater management, environmental engineering and environmental impact risk assessment.

Contents

Acknowledgment and How to Cite

The editor and publisher thank each of the authors who contributed to this book. The chapters in this book were previously published in various places in various formats. To cite the work contained in this book and to view the individual permissions, please refer to the citation at the beginning of each chapter. Each chapter was read individually and carefully selected by the editor; the result is a book that provides a nuanced look at pollution and the atmosphere. The chapters included are broken into three sections, which describe the following topics:

- Chapter 1 provides an overview of trends in emissions, air concentrations, and atmospheric depositions of heavy metals in Italy and of the relevant EU legislation and its goals, including directives on paints, batteries,and industrial emissions.
- Chapter 2 evaluates aspects related to the influence of raw material transport on the primary emissions of particulate matter from two plants in which thermal processes take place: a sintering plant and a residual municipal solid waste incineration plant.
- Chapter 3 provides a critical analysis of some processes determining a release of pollutants into the atmosphere, from residual municipal solid waste.
- Chapter 4 studies soot deposits on in-cylinder engine components and lubricating oil contamination when using biodiesel obtained from single-vegetable oil and its blends.
- Since agricultural machinery is an important source of emission of air pollutant in rural locations, chapter 5 offers an important examination of the effects of types of tractors and operation conditions on engine emission.
- Chapter 6 looks at recent engineering that reduce emissions and improve vehicle efficiency by using different sensors such as oxygen and mono-nitrogen oxides sensors, and utilizing exhaust gas recirculation to lower the oxygen concentration and reduce the temperature within the engines.
- Chapter 7 compares the life-cycle greenhouse gas intensities per megawatt-hour of electricity produced for a range of Australian and other energy sources, including coal, conventional liquefied natural gas, coal seam gas, and nuclear and renewables for the Australian export market.

- Chapter 8 assesses different fuel combinations that can be adopted to reduce the level of air pollution and greenhouse-gas emissions associated with energy generation are assessed.
- Chapter 9 evaluates the greenhouse gas footprint of natural gas obtained by high-volume hydraulic fracturing from shale formations, focusing on methane emissions.
- Chapter 10 is a life-cycle inventory analysis of household waste management scenarios for Kyoto with a special emphasis on food waste reduction activities.
- Chapter 11 uses Earth's measured energy imbalance, paleoclimate data, and simple representations of the global carbon cycle and temperature to define emission reductions needed to stabilize climate and avoid potentially disastrous impacts on the future of our planet.

List of Contributors

Frank Ackerman
Synapse Energy Economics, Cambridge, Massachusetts, United States of America

Hashem Akbari
Lawrence Berkeley National Laboratory

Kalyan Annamalai
Paul Pepper Professor of Mechanical Engineering, MEOB 307, Texas A&M University, College Station, Texas, USA

Johann Antoine
International Centre for Environmental and Nuclear Sciences, University of the West Indies, Kingston, Jamaica

Tiberiu Apostol
Prof., Power Engineering Faculty, University POLITEHNICA of Bucharest, Romania

Adrian Badea
Prof. Power Engineering Faculty, University POLITEHNICA of Bucharest, Romania

David J. Beerling
Department of Animal and Plant Sciences, University of Sheffield, Sheffield, South Yorkshire, United Kingdom

M. Catrambone
CNR-Institute of Atmospheric Pollution Research, Italy

G. Cattani
ISPRA, Italian National Institute for Environmental Protection and Research, Italy

Alessandro Chistè
M. Sc., Engineering Faculty Civil and Environmental Dep., University of Trento, Italy

Simona Ciuta
PhD, Power Engineering Faculty, University POLITEHNICA of Bucharest, Romania

Tom S. Clark
Principal Consultant (Carbon and Sustainability Consulting) WorleyParsons/Level 7, 250 St Georges Terrace, Perth 6000 Western Australia, Australia

Jaliliantabar Farzad
M.Sc. Students of Mechanics of Agricultural Machinery Department, Razi University of Kermanshah, Iran

A. Fino
CNR-Institute of Atmospheric Pollution Research, Italy

Charles Grant
International Centre for Environmental and Nuclear Sciences, University of the West Indies, Kingston, Jamaica

James Hansen
Earth Institute, Columbia University, New York, New York, United States of America

Paul E. Hardisty
Global Director, EcoNomics™ & Sustainability, WorleyParsons/Level 7, 250 St Georges Terrace, Perth 6000, Western Australia, Australia and Visiting Professor, Department of Civil and Environmental Engineering, Imperial College, London/Exhibition Road, South Kensington, London SW7 2AZ, UK

Paul J. Hearty
Department of Environmental Studies, University of North Carolina, Wilmington, North Carolina, United States of America

Rabbani Hekmat
Assistant professor of Mechanics of Agricultural Machinery Department, Razi University of Kermanshah, Iran

Yasuhiro Hirai
Environmental Preservation Research Center, Kyoto University

Ove Hoegh-Guldberg
Global Change Institute, University of Queensland, St. Lucia, Queensland, Australia

Robert W. Howarth
Department of Ecology and Evolutionary Biology, Cornell University, Ithaca, NY 14853

Shi-Ling Hsu
College of Law, Florida State University, Tallahassee, Florida, United States of America

Robert G. Hynes
Principal Consultant (Carbon and Sustainability Consulting), WorleyParsons/Level 10, 141 Walker Street, North Sydney 2000, New South Wales, Australia

Gabriela Ionescu
Dr., Dept. of Energy Production and Use, POLITEHNICA University of Bucharest, Romania

Anthony Ingraffea
School of Civil and Environmental Engineering, Cornell University, Ithaca, NY 14853, USA

Snehal S. Jadav
Agricultural & Food Engineering Department, Indian Institute of Technology

Prakash C. Jena
Agricultural & Food Engineering Department, Indian Institute of Technology

Nnenesi Kgabi
Department of Civil and Environmental Engineering, Polytechnic of Namibia, Windhoek, Namibia

Pushker Kharecha
Earth Institute, Columbia University, New York, New York, United States of America and Goddard Institute for Space Studies, NASA, New York, New York, United States of America

Ronnen Levinson
Lawrence Berkeley National Laboratory

Valerie Masson-Delmotte
Institut Pierre Simon Laplace, Laboratoire des Sciences du Climat et de l'Environnement (CEA-CNRS-UVSQ), Gif-sur-Yvette, France

Takeshi Matsuda
Environmental Preservation Research Center, Kyoto University

Lorestani Ali Nejat
Assistant professor of Mechanics of Agricultural Machinery Department, Razi University of Kermanshah, Iran

Camille Parmesan
Marine Institute, Plymouth University, Plymouth, Devon, United Kingdom and Integrative Biology, University of Texas, Austin, Texas, United States of America

Javadikia Payam
Assistant professor of Mechanics of Agricultural Machinery Department, Razi University of Kermanshah, Iran

N. Pirrone
Director of CNR-Institute of Atmospheric Pollution Research, Italy

Elena Cristina Rada
Ass. Researcher, Engineering Faculty, Civil and Environmental Dep., University of Trento, Italy

Marco Ragazzi
Associate Prof., Engineering Faculty, Civil and Environmental Dep., University of Trento, Italy

Hifjur Raheman
Agricultural & Food Engineering Department, Indian Institute of Technology

Gholami Rashid
M.Sc. Students of Mechanics of Agricultural Machinery Department, Razi University of Kermanshah, Iran

Johan Rockstrom
Stockholm Resilience Center, Stockholm University, Stockholm, Sweden

Eelco J. Rohling
School of Ocean and Earth Science, University of Southampton, Southampton, Hampshire, United Kingdom and Research School of Earth Sciences, Australian National University, Canberra, ACT, Australia

Jeffrey Sachs
Earth Institute, Columbia University, New York, New York, United States of America

Shin-ichi Sakai
Environmental Preservation Research Center, Kyoto University

Renee Santoro
Department of Ecology and Evolutionary Biology, Cornell University, Ithaca, NY 14853

Makiko Sato
Earth Institute, Columbia University, New York, New York, United States of America

Marco Schiavon
M. Sc., Engineering Faculty Civil and Environmental Dep., University of Trento, Italy

Pete Smith
University of Aberdeen, Aberdeen, Scotland, United Kingdom

Konrad Steffen
Swiss Federal Institute of Technology, Swiss Federal Research Institute WSL, Zurich, Switzerland

M. Strincone
CNR-Institute of Atmospheric Pollution Research, Italy

Werner Tirler
Dr., Eco-Research Srl Bolzano, Italy

Marco Tubino
Prof., Engineering Faculty, Civil and Environmental Dep., University of Trento, Italy

Lise Van Susteren
Center for Health and the Global Environment, Advisory Board, Harvard School of Public Health, Boston, Massachusetts, United States of America

Karina von Schuckmann
L'Institut Francais de Recherche pour l'Exploitation de la Mer, Ifremer, Toulon, France

Junya Yano
Environmental Preservation Research Center, Kyoto University

James C. Zachos
Earth and Planetary Science, University of California, Santa Cruz, CA, United States of America

Dino Zardi
Prof., Dr., Dept. of Civil and Environmental Engineering, University of Trento, Trento, Italy

Introduction

Air quality has a significant impact on human health. Assessing air pollution in complex morphologies has become an important issue in order to implement mitigation measures and limit emissions from the most relevant sources, such as waste incineration, traffic emissions, emissions from fuel and electricity production, and household emissions. These pollutants result in adverse health effects, material damage, damage to ecosystems, and global climate change.

Accurate assessments are the foundation for effective mitigation measures. Addressing these impacts requires an understanding of both the emission sources and the processes that transform the emissions in the atmosphere. For example, emissions from diverse sources contribute to the photochemical formation of secondary pollutants, such as tropospheric ozone and secondary particulate matter, with established impacts on health, livelihoods and the climate.

Air pollution engineering consists of two major components: (1) air pollution control and (2) air-quality engineering. Air pollution control focuses on the fundamentals of air pollutant formation in process technologies and the identification of options for mitigating or preventing air pollutant emissions. Air quality engineering deals with large-scale, multi-source control strategies, with focus on the physics and chemistry of pollutant interactions in the atmosphere. The articles in this compendium contain recent research in both areas.

—Marco Ragazzi

Heavy metals can cause adverse effects to humans, animals and ecosystems due to their bioavailability and toxicity in various environmental compartments. In the last decades, many policy strategies and measures have been taken at global, regional and local level in relation to heavy metals, due to their adverse effects and ability to be transported over long distances. Several EU measures have been adopted in order to control the pollution from heavy metals in the main sectors. Chapter 1, by Strincone and colleagues, will provide an overview of trends of emissions, air concentrations and atmospheric depositions of heavy metals in Italy and of the main relevant EU legislation and its goals (Directives on paints, batteries, industrial emissions, etc.) together with policies adopted at Italian level.

Chapter 2, by Ciuta and colleagues, presents some preliminary considerations on the role of direct particulate matter emissions vs induced transport emissions for two kinds of plants: incinerator and sintering plant. The developed balances demonstrate that in terms of total amount emitted, the emissions from not optimized transport of raw materials are comparable with the ones from the stacks of the sintering plant. That means it is important to promote initiatives for the adoption of modern engines in the transport system.

Chapter 3, by Ionescu and colleagues, provides a critical analysis of some processes determining a release of pollutants into the atmosphere, from residual municipal solid waste. The reference context is the situation in the North-East region of Italy. The role of waste composition and the implications of the available technologies are examined as well. Then the processes governing the dispersion into the atmosphere are discussed. Finally the different processes, by which emissions may determine an increased risk for human health, as well as criteria to be adopted in order to minimize this impact, are discussed.

In Chapter 4, Raheman and colleagues evaluated A 10.3-kW single-cylinder water-cooled direct-injection diesel engine using blends of biodiesel (B10 and B20) obtained from a mixture of mahua and simarouba oils (50:50) with high-speed diesel (HSD) in terms of brake specific fuel consumption, brake thermal efficiency, and exhaust gas temperature and emissions such as CO, HC, and NOx. Based on performance and emissions, blend B10 was selected for long-term use. Experiments were also conducted to assess soot deposits on engine components, such as cylinder head, piston crown, and fuel injector tip, and addition of wear metal in the lubricating oil of diesel engine when operated with the biodiesel blend (B10) for 100 h. The amount of soot deposits on the engine components was found to be, on average, 21% lesser for B10-fueled engine as compared with HSD-fueled engine due to better combustion. The addition of wear metals such as copper, zinc, iron, nickel, lead, magnesium, and aluminum, except for manganese, in the lubricating oil of B10-fueled engine after 100 h of engine operation was found to be 11% to 50% lesser than those of the HSD-fueled engine due to additional lubricity.

Agricultural machinery is an important source of emission of air pollutant in rural locations. Chapter 5, by Rashid and colleagues, deals with the effects of types of tractors and operation conditions on engine emission. The values of some exhaust gases (HC, CO, CO_2, O_2 and NO) from two common tractors (MF285 and U650) at three situations (use of ditcher, plowing and cultivator) were evaluated in the West of Iran (Kermanshah). In addition, engine oil temperature at operation conditions was measured. Also results showed the values of exhaust HC and O_2 of MF285 are lower than U650, while the other exhausts gases (CO, CO_2, and NO) of MF285 are higher than U650. Value of NO emission increased as engine oil temperature increased. All of exhaust gases except CO have a significant relationship with type of tractors, while all of measured gases have a significant relationship with installed instruments at 1%.

Development and economic growth throughout the world will result in increased demand for energy. Currently almost 90% of the total world energy demand is met utilizing fossil fuels [1]. Petroleum and other liquid fuels include 37% of the total fossil reserves consumed for transportation and other industrial processes [1]. Emission of harmful gases in the form of nitrogen oxides, sulphur oxides and mercury are the major concerns from the combustion of conventional energy sources. In addition to these pollutants, huge amount of carbon dioxide is liberated into the atmosphere. Carbon dioxide is one of the green house gases which cause global warming. Chapter 6, by Annamalai, argues that though technology is being developed to sequester the CO_2 from stationary power generating sources, it is difficult to implement such a technology in non-stationary automobile IC engines.

Electricity generation is one of the major contributors to global greenhouse gas emissions. Transitioning the world's energy economy to a lower carbon future will require significant investment in a variety of cleaner technologies, including renewables and nuclear power. In the short term, improving the efficiency of fossil fuel combustion in energy generation can provide an important contribution. Availability of life cycle GHG intensity data will allow decision-makers to move away from overly simplistic assertions about the relative merits of certain fuels, and focus on the complete picture, especially the critical roles of technology selection and application of best practice. Chapter 7, by Hardisty and colleagues, compares the life-cycle greenhouse gas (GHG) intensities per megawatt-hour (MWh) of electricity produced for a range of Australian and other energy sources, including coal, conventional liquefied natural gas (LNG), coal seam gas LNG, nuclear and renewables, for the Australian export market. When Australian fossil fuels are exported to China, life cycle greenhouse gas emission intensity in electricity production depends to a significant degree on the technology used in combustion. LNG in general is less GHG intensive than black coal, but the gap is smaller for gas combusted in open cycle gas turbine plant (OCGT) and for LNG derived from coal seam gas (CSG). On average, conventional LNG burned in a conventional OCGT plant is approximately 38% less GHG intensive over its life cycle than black coal burned in a sub-critical plant, per MWh of electricity produced. However, if OCGT LNG combustion is compared to the most efficient new ultra-supercritical coal power, the GHG intensity gap narrows considerably. Coal seam gas LNG is approximately 13–20% more GHG intensive across its life cycle, on a like-for like basis, than conventional LNG. Upstream fugitive emissions from CSG (assuming best practice gas extraction techniques) do not materially alter the life cycle GHG intensity rankings, such is the dominance of end-use combustion, but application of the most recent estimates of the 20-year global warming potential (GWP) increases the contribution of fugitives considerably if best practice fugitives management is not assumed. However, if methane leakage approaches the elevated levels recently reported in some US gas fields (circa 4% of gas production) and assuming a 20-year methane GWP, the GHG intensity of CSG-LNG generation is on a par with sub-critical coal-fired generation. The importance of applying best practice to fugitives management in Australia's emerging natural gas industry is evident. When exported to China for electricity production, LNG was found to be 22–36 times more GHG intensive than wind and concentrated solar thermal (CST) power and 13–21 times more GHG intensive than nuclear power which, even in the post-Fukushima world, continues to be a key option for global GHG reduction.

In Chapter 8, Kgabi and colleagues assess the different fuel combinations that can be adopted to reduce the level of air pollution and GHG emissions associated with the energy generation; and the air pollution and global warming effects of the Jamaican electricity generation fuel mix are determined. Based on the energy production and consumption patterns, and global warming potentials, the authors conclude that: an increase in energy consumption and production yields an increase in GHGs and other major pollutants; choice of the fuel mix determines the success of GHG emissions reductions; and there is no single fuel that is not associated with GHG or other air pollution or environmental degradation implications.

In April 2011, Howarth and colleagues—the authors of Chapter 9—published the first comprehensive analysis of greenhouse gas (GHG) emissions from shale gas obtained by hydraulic fracturing, with a focus on methane emissions. Their analysis was challenged by Cathles et al. (2012). Here, they respond to those criticisms. The authors stand by their approach and findings. The latest EPA estimate for methane emissions from shale gas falls within the range of our estimates but not those of Cathles et al. which are substantially lower. Cathles et al. believe the focus should be just on electricity generation, and the global warming potential of methane should be considered only on a 100-year time scale. The authors' analysis covered both electricity (30% of US usage) and heat generation (the largest usage), and they evaluated both 20- and 100-year integrated time frames for methane. Both time frames are important, but the decadal scale is critical, given the urgent need to avoid climate-system tipping points. Using all available information and the latest climate science, they conclude that for most uses, the GHG footprint of shale gas is greater than that of other fossil fuels on time scales of up to 100 years. When used to generate electricity, the shale-gas footprint is still significantly greater than that of coal at decadal time scales but is less at the century scale. The authors reiterate our conclusion from our April 2011 paper that shale gas is not a suitable bridge fuel for the 21st Century.

Source-separated collection of food waste has been reported to reduce the amount of household waste in several cities including Kyoto, Japan. Food waste can be reduced by various activities including preventing edible food loss, draining moisture, and home composting. These activities have different potentials for greenhouse gas (GHG) reduction. Therefore, in Chapter 10 Matsuda and colleagues conducted a life-cycle inventory analysis of household waste management scenarios for Kyoto with a special emphasis on food waste reduction activities. The primary functional unit of this study was "annual management of household combustible waste in Kyoto, Japan." Although some life-cycle assessment scenarios included food waste reduction measures, all of the scenarios had an identical secondary functional unit, "annual food ingestion (mass and composition) by the residents of Kyoto, Japan." The authors analyzed a typical incineration scenario (Inc) and two anaerobic digestion (dry thermophilic facilities) scenarios involving either source-separated collection (SepBio) or nonseparated collection followed by mechanical sorting (MecBio). They assumed that the biogas from anaerobic digestion was used for power generation. In addition, to evaluate the effects of waste reduction combined with separate collection, three food waste reduction cases were considered in the SepBio scenario: (1) preventing loss of edible food (PrevLoss); (2) draining moisture contents (ReducDrain); and (3) home composting (ReducHcom). In these three cases, they assumed that the household waste was reduced by 5%. The GHG emissions from the Inc, MecBio, and SepBio scenarios were 123.3, 119.5, and 118.6 Gg CO_2-eq/year, respectively. Compared with the SepBio scenario without food waste reduction, the PrevLoss and ReducDrain cases reduced the GHG emissions by 17.1 and 0.5 Gg CO_2-eq/year. In contrast, the ReducHcom case increased the GHG emissions by 2.1 Gg CO_2-eq/year. This is because the biogas power production decreased due to the reduction in food waste, while the electricity consumption increased in response to home composting. Sensitivity analyses revealed that a reduction of only 1% of the household waste by food loss prevention has the same GHG reduction effect as a 31-point increase (from 50% to 81%) in the food waste separation rate. The authors found that prevention of food losses enhanced by separate collection led to a significant reduction in GHG emissions. These findings will be useful in future studies designed to develop strategies for further reductions in GHG emissions.

In Chapter 11, Hansen and colleagues assess climate impacts of global warming using ongoing observations and paleoclimate data. We use

Earth's measured energy imbalance, paleoclimate data, and simple representations of the global carbon cycle and temperature to define emission reductions needed to stabilize climate and avoid potentially disastrous impacts on today's young people, future generations, and nature. A cumulative industrial-era limit of ~500 GtC fossil fuel emissions and 100 GtC storage in the biosphere and soil would keep climate close to the Holocene range to which humanity and other species are adapted. Cumulative emissions of ~1000 GtC, sometimes associated with 2°C global warming, would spur "slow" feedbacks and eventual warming of 3–4°C with disastrous consequences. Rapid emissions reduction is required to restore Earth's energy balance and avoid ocean heat uptake that would practically guarantee irreversible effects. Continuation of high fossil fuel emissions, given current knowledge of the consequences, would be an act of extraordinary witting intergenerational injustice. Responsible policymaking requires a rising price on carbon emissions that would preclude emissions from most remaining coal and unconventional fossil fuels and phase down emissions from conventional fossil fuels.

PART I

INTRODUCTION:
WHERE WE CAME FROM
AND WHERE WE'RE HEADED

CHAPTER 1

Emissions, Air Concentrations and Atmospheric Depositions of Arsenic, Cadmium, Lead, and Nickel in Italy in the Last Two Decades: A Review of Recent Trends in Relation to Policy Strategies Adopted Locally, Regionally, and Globally

M. STRINCONE, A. FINO, G. CATTANI, M. CATRAMBONE, AND N. PIRRONE

1.1 INTRODUCTION

According to the international scientific literature, heavy metals can cause adverse effects to humans, animals and ecosystems due to their bioavailability and toxicity in various environmental compartments. In the last decades, many policy strategies and measures have been taken at global, regional and local level in relation to heavy metals, in particular

Emissions, Air Concentrations and Atmospheric Depositions of Arsenic, Cadmium, Lead, and Nickel in Italy in the Last Two Decades: A Review of Recent Trends in Relation to Policy Strategies Adopted Locally, Regionally, and Globally. © Strincone M, Fino A, Cattani G, Catrambone M, and Pirrone N. E3S Web of Conferences *1* (2013), DOI: 10.1051/e3sconf/20130138003. Licensed under Creative Commons Attribution 2.0 Generic License, http://creativecommons.org/licenses/by/2.0.

cadmium and lead, due to their adverse effects and ability to be transported long distances.

The first international legally binding instrument to deal with problems of air pollution on a broad regional basis is represented by the 1979 Geneva Convention on Long Range Transboundary Air Pollution (LRTAP), which has been extended by eight specific protocols. Following this treaty, further measures have been adopted both at regional level, for example the European Union strategies on air quality, mercury and waste treatment, and at global level, for example the discussion in progress under the UNEP umbrella dealing with lead and cadmium pollution.

1.2 EMISSIONS IN ITALY AND IN EUROPE

Lead and cadmium occur in the environment as a result of both natural releases and releases associated with human activities.

The major natural emissions of lead come from volcanoes, airborne soil particles, sea spray, biogenic material and forest fires, while the major anthropogenic contribution to emissions is due to the extraction of metal residues and of other minerals like coal and lime, industrial activities, smelters and metal/oil refineries (UNEP,2010a).

The main sources of cadmium anthropogenic contamination is associated with mining, metallurgical industries, the use of fertilizers containing phosphates from mineral products, of paint and coating industries and electroplating industries (UNEP,2010a). In Europe major anthropogenic sources of nickel are stationary combustion (55 %) and mobile sources and machinery other than road transport (30 %). With a view to air quality the relevant sources are petroleum refining and fugitive emissions from the electric arc furnace steel works. Important natural sources of nickel are windblown soil and volcanoes. Anthropogenic sources considerably outweigh natural sources. In Europe 86 % of total arsenic emissions in the in 1990 was emitted by stationary combustion. However, in general the emissions from this sector do not result in relevant ambient air concentrations as they are released via sufficiently high stacks. In general anthropogenic sources outweigh natural sources; the global natural share is estimated at 25 %, mainly from volcanoes. On a local scale there may be more significant contributions up to 60 % from weathering processes in regions rich in sulphidic ore deposits (European Commission, 2003).

Emissions in Italy can be evaluated considering the data reported under the relevant EU regulations and the UNECE Protocol on heavy metals under the Convention on long range transboundary air pollution.

In particular the emissions into air and water from the major industrial plants are listed under the EPER Register, replaced by the E-PRTR register from 2007.

The UNECE Protocol on heavy metals foresees that Each Party shall develop and maintain emission inventories for several heavy metals, including lead, cadmium, arsenic and nickel. Italy has submitted the time series from 1990 to 2010 to the Convention Secretariat (ISPRA).

At national level, the Italian Institute of Statistics reports on heavy metals emissions from specific sectors for the period 1990-2007 (ISTAT).

According to the available data, the main emitting sectors for cadmium and lead include power production, waste treatment and manufacturing activities, with particular reference to the production of metals, metal products, non metal minerals and chemicals.

In addition, the farming sector has a considerable responsibility for cadmium emissions, while the emissions of lead from transport, storage and communications are and have been significant, especially in the past.

The figures (Figures 1 and 2) show the quantity of emissions of cadmium, lead, arsenic and nickel in 2007.

1.3 AIR CONCENTRATIONS IN ITALY AND IN EUROPE

The Directive 2004/107/EC (European Parliament, 2004) sets a target value for arsenic in ambient air concentration of 6 ng/m^3 to be calculated on the total content in the PM_{10} fraction, averaged over a calendar year. According to air quality data available at Italian level (Source: Regions, Autonomous Provinces, Regional Environmental Protection Agencies, National Research Council, National Institute for Environmental Protection and Research and National Institutes of Health) and European level (Source: Airbase) urban levels of arsenic are generally below the target value of 6 ng/m^3 showing a range of 0,5 to 5,5 ng/m^3. At European level urban background levels show a range of 0.5 to 3 ng/m³. Arsenic concentrations monitored near industrial installations may be up to one order of magnitude higher depending on the type of installation and the distance and position of the monitoring site (European Commission, 2003).

According to scientific literature and to air quality data exchanged amongst Member States in Europe, urban background levels of cadmium show a range of 0.2 to 2.5 ng/m³. Cadmium concentrations near industrial installations may be higher, depending on the type of installation and the distance and position of the monitoring site (European Commission, 2003). According to air quality data available at Italian level (Source: Regions, Autonomous Provinces, Regional Environmental Protection Agencies, National Research Council, National Institute for Environmental Protection and Research and National Institutes of Health) and European level (Source: Airbase) urban traffic and background levels are below the target value of 5 ng/m3 established by the so-called Fourth Daughter Directive, the Directive 2004/107/EC (European Parliament, 2004) showing a range of 0,1 to 3,0 ng/m³.

According to scientific literature and to air quality data exchanged amongst Member States in Europe, urban background levels of lead in Italy are below the limit value established by Directive 99/30/EC (Council of European Union, 1999) and currently in force according to Directive 2008/50/EC (European Parliament, 2008) of 0.5 µg/m³ expressed as an average yearly concentration for the protection of human health. Lead emissions have been reduced considerably as a result of the use of unleaded gasoline. Owing to decreases in the lead content of gasoline, there has been a trend towards lower air lead values both at European and Italian levels in the last years.

Referring to nickel the Directive 2004/107/EC (European Parliament, 2004) provides a target value of 20 ng/m³ for the total content in the PM_{10} fraction, averaged over a calendar year.. According to air quality data available at Italian level (Source: Regions, Autonomous Provinces, Regional Environmental Protection Agencies, National Institute for Environmental Protection and Research) and European level (Source: Airbase) urban levels are below the target value of 20 ng/m³ showing a range of 5 to 15 ng/m³. Nickel concentrations monitored near industrial installations may be up to one order of magnitude higher depending on the type of installation and the distance and position of the monitoring site.

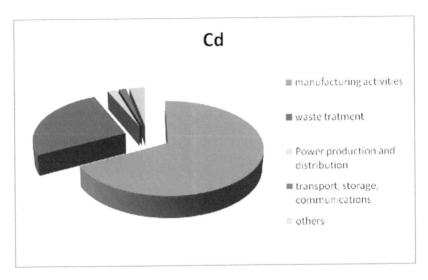

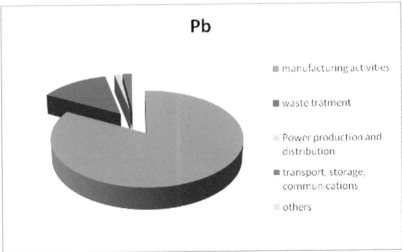

FIGURE 1: The quantity of emissions of cadmium (top) and lead (bottom) in 2007, according to the ISTAT.

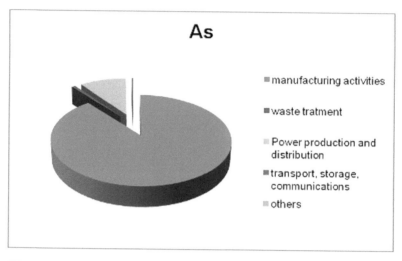

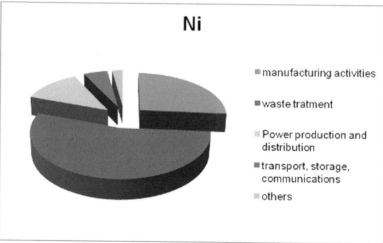

FIGURE 2 The quantity of emissions of arsenic (top) and nickel (bottom) in 2007, according to the ISTAT.

1.4 ATMOSPHERIC DEPOSITIONS IN ITALY AND IN EUROPE

The European Monitoring and Evaluation Programme (EMEP) is a scientifically based and policy driven programme under the Convention on Long-range Transboundary Air Pollution for international cooperation to solve transboundary air pollution problems. In particular, one of the activity of EMEP is to collect information on modelled and measured concentrations and deposition of lead, cadmium and mercury, their transboundary transport and atmospheric load to regional seas.

The EMEP has four assigned centers that annually make available reports on, among others, the transboundary pollution of the environment due to lead, cadmium and mercury, which are considered in this paper (EMEP).

1.5 COMPARISON OF TRENDS IN ITALY AND IN EUROPE

Considering the emission trends of cadmium, lead, arsenic and nickel in Italy, there has been a decrease of emissions in the last two decades.

For the example, the emissions of cadmium, lead and nickel in 1990 from manufacturing activities were 8.2, 397 and 52.3 tons respectively, while the emissions from the same sectors were 5.2, 230 and 26.6 tons respectively in 2007.

In addition the lead emission from the wholesale and retail trade, repair of motor vehicles, motorcycles and personal and household goods which was 807.5t in 1990 and 32 kg in 2007.

The lead emission from the transport, storage and communications sector were 373.3 t in 1990 and 5.4 t in 2007.

In this paper air emissions, concentrations and depositions measurements are compared considering the Italian and European situation (Pacyna et al, 2007), (Pacyna et al., 2009), (Storch et al., 2003), (EMEP).

1.6 POLICY STRATEGIES AT GLOBAL LEVEL AND THEIR EFFECTS

The Protocol on heavy metals under the UNECE LRTAP Convention states each Party shall reduce its total annual heavy metals air emissions of each heavy metal (listed in annex I) taking effective measures, appropriate to specific circumstances.

The Protocol on Heavy Metals sets legally binding limit values for the emission of particulate of 10 mg/m^3 from hazardous and medical waste incineration, which reduce emissions of heavy metals indirectly. It sets also emission limit values on the emission of mercury to 0,05 mg/m^3 from hazardous waste incineration and 0,08 mg/m^3 from municipal waste incineration (UNECE).

Considering the concerns related to the adverse effects to humans and the environment caused by lead and cadmium at global level, UNEP has started a collection of information on these two metals.

In relation to this, the final reviews of scientific information on lead and cadmium with an overview of existing and future national actions, (including legislation relevant to lead and cadmium), have been presented at the 26th GC/GMEF of UNEP in 2011. UNEP has identified three current priorities for action in connection with lead and cadmium, which are lead in paint, lead in fuels and in lead and cadmium batteries (UNEPc).

1.7 POLICY STRATEGIES AT EUROPEAN LEVEL AND THEIR EFFECTS

Several EU measures have been adopted in order to control the pollution from lead, cadmium, arsenic and nickel in the main sectors.

The poster will list the main relevant EU legislation and its goal (Directives on paints, batteries, industrial emissions, etc.)

1.8 POLICY STRATEGIES AT ITALIAN LEVEL AND THEIR EFFECTS

At national level, in addition to the legislative acts to transpose the UE directives, the policy makers, scientists and stakeholders have started the definition of a national plan in order to reduce caused by particulate matter and heavy metals.

It will be underlined main relevant strategies adopted at Italian level and possible measures for future management of heavy metal pollution.

REFERENCES

1. European Commission: Position paper on Lead, 1997.
2. European Commission: Ambient air pollution by AS, Cd and Ni compounds. Position Paper, Office For Official Publications of The European Communities, 1- 315, October 2000.
3. European Commission: Proposal for a Directive of the European Parliament and of the Council relating to arsenic, cadmium, mercury, nickel and polycyclic aromatic hydrocarbons in ambient air, COM(2003) 423 final, 2003.
4. European Parliament: Directive 2004/107/EC of the European Parliament and of the Council of 15 December 2004 relating to arsenic, cadmium, mercury, nickel and polycyclic aromatic hydrocarbons in ambient air, Official Journal of the European Union, L 23, 3-16, 26.1.2005.
5. European Parliament: Directive 2008/50/EC of the European Parliament and of the Council of 21 May 2008 on ambient air quality and cleaner air for Europe, Official Journal of the European Union, L 152, 2008.
6. European Union: Council Directive 1999/30/EC of 22 April 1999 relating to limit values for sulphur dioxide, nitrogen dioxide and oxides of nitrogen, particulate matter and lead in ambient air, Official Journal of the European Communities, L 163, 1999.
7. Pacyna, J. M., Pacyna, E. G., Aas, W.: Changes of emissions and atmospheric deposition of mercury, lead, and cadmium, Atmospheric Environment 43, 117–127, 2009.
8. Storch, V., Cabrala, M. C.-, Hagnera, C., Fesera, F., Pacyna, J., Pacyna, E., Kolb, S.: Four decades of gasoline lead emissions and control policies in Europe: a retrospective assessment, The Science of the Total Environment 311, 151–176, 2003.
9. ISPRA_The Institute for Environmental Protection and Research, (http://www.sinanet.isprambiente.it/it/sinanet/sstoriche)
10. ISTAT_ National Institute of Statistics (http://www3.istat.it/salastampa/comunicati/non_calendario/20100202_00/).
11. UNEP, chemicals branch, DTIE. Final review of scientific information on cadmium. December 2010a.
12. UNEP, chemicals branch, DTIE. Final review of scientific information on lead. December 2010b.
13. UNEP_Lead and Cadmium _Priorities for actions (http://www.unep.org/hazardoussubstances/LeadCadmium/PrioritiesforAction/tabid/6171/Default.aspx)
14. Pacyna, E. G.., Pacyna, J. M., Fudalac, J.,. StrzeleckaJastrzabc, E, Hlawiczkac, S., Panasiukd, D., Nittere, S., Preggere, T., Pfeiffere, H., Friedrich, R.: Current and

future emissions of selected heavy metals to the atmosphere from anthropogenic sources in Europe.Atmospheric Environment 2007; 4:18557– 8566.

15. UNECE Protocol on heavy metals (http://www.unece.org/fileadmin/DAM/env/lr-tap/full%20text/1998.Heavy.Metals.e.pdf).

16. EMEP_ European Monitoring and Evaluation Programme (http://www.emep.int/).

PART II

WASTE INCINERATION

CHAPTER 2

Role of Feedstock Transport in the Balance of Primary PM Emissions in Two Case-Studies: RMSW Incineration vs. Sintering Plant

SIMONA CIUTA, MARCO SCHIAVON, ALESSANDRO CHISTÈ, MARCO RAGAZZI, ELENA CRISTINA RADA, MARCO TUBINO, ADRIAN BADEA, AND TIBERIU APOSTOL

2.1 INTRODUCTION

One of the main air quality indicators is the particulate matter (PM) concentration at ground level. It is demonstrated that small aerosol particles or particulate matter (as PM_{10} and $PM_{2.5}$) affect air quality and can have significant effects on human's health [1]. Anomalous exposure to PM can shorten life expectancy, hospital admissions and emergency room visits. For these reasons, various national and international institutions [2,3] have established regulations to reduce PM concentration caused by human activities and to set adequate PM concentration limits.

Role of Feedstock Transport in the Balance of Primary PM Emissions in Two Case-Studies: RMSW Incineration vs. Sintering Plant. © Ciuta S, Schiavon M, Chistè A, Ragazzi M, Rada EC, Tubino M, Badea A, and Apostol T. UPB Scientific Bulletin, Series D: Mechanical Engineering 74,1 (2012). http://www.scientificbulletin.upb.ro/rev_docs_arhiva/rezca2_584373.pdf. Reprinted with permission from the authors and publishers.

Assessing air pollution in complex morphologies becomes an important issue in order to implement mitigation measures and limit emissions from the most relevant sources, such as traffic, manufacturing activities, heating and energy production. One of the consequences of the climate is the thermal stratification of the atmosphere within the valley, which makes the dilution of pollutants difficult [4].

The aim of this study is to evaluate preliminarily some aspects related to the influence of raw material transport on the primary emissions of PM from two plants in which thermal processes take place: a sintering plant and a Residual Municipal Solid Waste (RMSW) incineration plant. In this paper, this two industrial plants are supposed to be situated in a valley in the North of Italy. The total population of the virtual case-study Province is 519,800 [5]. Generated data are not referable to existing and proposed plants.

2.2 MATERIALS AND METHODS

The hypothesized incineration plant will treat 103,000 t y^{-1} of RMSW mainly. The efficiency of selective collection in the proposed case-study was supposed to reach the 65% of the total produced waste in 2013, taking into account an amount of 175 kgRMSW inh^{-1} y^{-1}. This plant will generate a maximum thermal power of 60 MW and will ensure a minimum net electrical efficiency of 23% [6]. Stack emissions have to comply with the limit values for the regulated pollutants and must be guaranteed lower than 2 mgPM Nm-3 for this case-study.

In order to obtain the primary emissions for the RMSW incineration plant, it can be considered a flow of waste (F) depending on the number of hours per year of operation. The energy potential of the material entering the incinerator can be related to its Lower Heating Value (LHV) and the inert content (A) can be assessed from the waste characteristics.

The emission factor for the particulate matter can be calculated using the following expression:

$$e = \frac{A(1-x)\cdot(1-y)}{LHV}\left[\frac{kg}{kJ}\right] \tag{1}$$

Where: x – degree of retention of ash in the outbreak, y – particulate matter retention efficiency, LHV – lower heating value

The total content of particulate matter emitted (c) can be determined taking into account the volume factor (F_v), which is defined as the ratio of total volume of flue gas and the amount of heat related to the fuel introduced into the boiler:

$$c = \frac{e}{F_v} \left[\frac{mg}{m_N^3} \right] \tag{2}$$

The amount of particulate matter can be determined taking into account the following expression:

$$PM = F \cdot LHV \cdot e \left[\frac{kg}{h} \right] \tag{3}$$

An alternative way can be adopted using a specific flow-rate that can be related to the LHV of the waste, the yearly amount of waste burnt and the concentration at the stack.

Concerning the virtual sintering plant, raw materials (530,000 t y^{-1}) for this plant are principally ferrous wastes which arrive through heavy vehicles from several points of the region. The final products are billets and bars of iron.

The emissions into the atmosphere from the plant can be primary or secondary. The first ones come from the raw material processing into the furnace and from the refining furnace. The second ones come from other operations into the plant (spillage, ladle transport operation, etc.) and are called diffuse emissions.

The emission treatment is characterized by two lines: one for the primary emissions and a part of the secondary, and one only for the secondary emissions. The two lines are connected with two different chimneys called L1 (first line) and L2 (second line).

Emission values of PM at the stacks and flow-rates are supposed to be available on-line allowing the assessment of the PM emission flows (expressed in mg h^{-1}).

For both the plants, in order to calculate the PM emissions related to the systems of road transportation, an emission model (COPERT 4) was used and adapted for this cases study. The COPERT 4 algorithm is part of the EMEP/CORINAIR emission inventory guidebook [7]. This methodology has been developed by EEA within the European Topic Centre on Air and Climate Change (ETC/ACC) activities, with the intention of providing a set of tools for the compilation of emission inventories to the European Countries [7].

The COPERT algorithm estimates emissions of all the main pollutants (CO, NOx, VOC, PM, NH_3, SO_2, heavy metals) as well as greenhouse gases (CO_2, N_2O, CH_4) [7]. These pollutants can be divided into four main groups:

- pollutants whose a detailed methodology for the calculation of the emission factors exists (CO, NOx, VOC, CH_4, PM);
- compounds whose the emission factors are calculated according to the fuel consumption, falling within the second group (CO_2, SO_2 and heavy metals);
- pollutants whose a simplified methodology is applied, since detailed studies are not available (NH_3, N_2O, PAHs, dioxins and furans);
- profiles of alkanes, alkenes, alkynes, ketones, aldehydes, aromatics and cycloalkanes, derived as fractions of the total NMVOCs [7].

For each pollutant, the algorithm calculates the emission factors (expressed in g km^{-1} $vehic^{-1}$) relative to specific vehicle classes which the vehicles belong to.

To apply the model, it was necessary to evaluate the vehicle fluxes for the two considered scenarios. A different approach was used to estimate the fluxes of heavy duty vehicles that deliver raw material to the plants.

The delivery of waste to the incineration plant will be provided by a system of road transportation, which will be based on the use of heavy vehicles. The typical journey of a vehicle starts in a collection centre, where the truck loads bulky waste, residual waste and scraps from the separate waste collection; later the truck moves to the incineration plant, unloads the waste and comes back empty to the original collection centre. The Province is divided into eleven districts (numbered from C1 to C11), plus two districts represented by the two main cities (C12 and C13). In the case of incineration plant, an analysis of the transportation system organization was proposed taking into account that almost all districts will have their own collection centers, where trucks load bulky and residual waste and move to the incineration plant. The frequency of journeys from each collection center depends on the amount of deposited waste, that has been evaluated on the basis of the catchment area of the districts, the estimated evolution of the population in the future years and the decrease of waste production.

Since all the routes between the collection centers and the incineration plant are long itineraries and extra-urban paths, the effects of slowdowns and accelerations (which are typical for urban routes) can be neglected as a first approximation. Consequently, the average speed approach was adopted. The slope effects were taken into account, since every route does not follow a flat pathway (excepted for districts C4, C5 and the district located in the bottom of the valley). Besides the road gradient itself, slope correction factors for heavy vehicles depend on the COPERT vehicle class, depending on the vehicle mass, since the classification for heavy vehicles is based on the gross weight. Hence, different load conditions lead to different correction factors. Moreover, slope correction factors for the same vehicle class are not merely equal in modulus and opposite in sign for a round journey. Consequently, when dealing with non-flat routes, slope effects should not be neglected.

In addition, in order to evaluate the positive effects of the latest emission standards on the decrease of the emitted pollutants, both EURO 1 and EURO 5 heavy vehicles were considered in this study. Since this calculation is based on the average speed approach, a mean speed of 50 km h^{-1} was adopted, both for the outward and for the return journey.

The transportation from the collection centers to the incinerator will take place by means of 26 tons heavy vehicles (with tare of 10 t and load capacity of 16t). For those districts without any collection centre, the transportation will be made by 12 tons heavy vehicles (with tare of 4 t and load capacity of 8 t). For these two districts the delivery of waste to the plant was assumed that will be carried out directly at the end of the collection service of RMSW from the bins located in the territory.

According to the COPERT classification, in the case of the sintering plant, the transportation vehicle classes which the assumed trucks belong to are vehicles with a weight between 16 t and 32 t (for the return journey). Considering the maximum mass of material the trucks can be loaded is 30 t (which is the difference between the gross weight and the tare of each vehicle) and since the amount of raw material entering the plant is 530,000 t y^{-1}, the number of journeys to transport raw materials to the sintering plant during one year is 17,667 according to this virtual scenario.

In the case of sintering plant, it was assumed that the trucks arrive to the sintering plant from three different points of the region, with different distances and road slopes.

- the first one is 24 km long with a positive slope of 3‰ (first path),
- the second one is 60 km long with a positive slope of 3‰ (second path),
- the third one is 89 km long with a positive slope of 7‰ (third path).

The slope is always positive for each path, this means that for the outward journey the road is uphill on average for each path. Moreover it was assumed that the 50% of the trucks follow the second path, 25% of them follow the first one and 25% the third one.

For the considered vehicle classes, COPERT provides the mass of PM emitted for unit of time (1 year in this case) and length (km). In this way, the PM emission values for each path and class were obtained, in term of kg y^{-1} km^{-1}. The calculated PM emission values were multiplied by the length of the respective routes and the total emission values obtained for the outward and the return journeys (expressed in kg y^{-1}) along each path were added up.

Similarly to what performed for the waste transportation system, this procedure was conducted considering both EURO 1 and EURO 5 vehicles. Moreover, in analogy with the previous case study, since the journeys take place on sloping routes, slope correction factors were introduced to consider the effect of road gradient on the emissions. Finally, since all routes to the sintering plant are long itineraries and extra-urban paths, the average speed approach was adopted, as for the case of the incineration plant.

To make a comparison between transportation and primary emissions from the sintering plant, it is necessary to obtain the number of heavy vehicles for the transport of raw material, the covered distance and the PM emission values from the stack of the plant.

2.5 RESULTS

In the case of the RMSW incineration plant, emissions from road transport are a small part of the total emissions, as the first ones remain at least one order of magnitude below the latter as explained below. The lowest emissions are achieved when considering EURO 5 vehicles instead of EURO 1. In fact, PM emissions of the latter are almost five times higher than those related to EURO 5 those for heavy vehicles greater than 32 t (for the outward journey) and the heavy trucks (Table 1). Considering stack emissions, assuming a specific flow-rate as 7.5 $Nm^3/kgRMSW$, an emission of PM of about 1,545 kg (or less) on yearly basis can be assessed from the second method of calculation (preferred thanks to its simplicity). That value confirms the difference in order of magnitudes between stack and transport emissions.

TABLE 1: Annual PM emission values related to the road transportation system of the incineration plant, calculated for the two scenarios (with EURO 1 and EURO 5 heavy vehicles)

District	Distance from the plant [km]	Average slope [%]	Mean annual mileage (outward journey only) [km y⁻¹]	Annual PM emissions [kg y⁻¹]	
				EURO 1 vehicles	EURO 5 vehicles
C1	70	1.1	7,749	6.769	1.636
C2	100	0.5	7,774	6.791	1.642
C3	44	0.5	7,285	6.364	1.538
C6 and C7	38	0.8	15,908	13.897	3.360
C8	45	1.0	19,370	16.921	4.091
C9	51	-0.2	28,276	28.349	4.664
C10 and C12	40	-0.1	27,255	27.325	4.496
C11	97	1.2	16,204	14.155	3.422
C4, C5 and C13	-	0.0	54,489	30.031	8.565
TOTAL				150.603	33.413

For the case of the sintering plant, average values of concentration and flow can be calculated starting from on-line data available in the sector: resulting data are presented in Table 2, in order to finally obtain PM emission values.

To complete the calculation, the number of working hours per year for the sintering plant is needed: it was supposed that plant works in the average 16.5 hours in a day (in the average) and 335 days in a year.

The average values of concentration, flow and the calculated annual PM emissions are presented in Table 2.

TABLE 2: Average values of PM concentration assumed at chimney level, flow-rate and PM emission from two chimneys

	L1	L2
PM concentration [mg Nm^{-3}]	0.37	0.19
Flow [Nm^3 h^{-1}]	553,856	691,350
PM emission [mg h^{-1}]	204,927	131,357
PM emission [kg y^{-1}]	1,133	726
Total PM emissions [kg y^{-1}]		1,859

The final results of the emission calculation related to the sintering plant transportation system are shown in Table 3. In case of old heavy vehicles the contribution of transportation can be comparable with the one of PM from the stacks.

2.6 CONCLUSIONS

In conclusion an important difference emerges between the emissions related to EURO 1 and EURO 5 trucks used to simulate the transportation system. Due to the technological progresses made in the last years, EURO 5 trucks emit twenty times less PM compared to EURO 1 trucks.

TABLE 3: Annual PM emission values related to the road transportation system of the sintering plant, calculated for the two scenarios (with EURO 1 and EURO 5 heavy vehicles)

		Annual PM emissions [kg y⁻¹]	
		EURO 1 vehicles	EURO 5 vehicles
First Path	Outward journey	57.63	2.93
	Return Journey	26.16	7.55
Second Path	Outward journey	288.15	14.63
	Return Journey	130.82	6.64
Third Path	Outward journey	213.71	10.85
	Return Journey	97.02	4.92
TOTAL		813.50	41.29

In case of sintering plants, EURO 1 trucks produce emissions comparable with those released from the chimneys. Hence, at regional scale, road transport can play an important role within the emissive balance. It must be noted that only transportation of metallic minerals were considered in this study. As obvious, several kinds of raw materials are needed for the production of steel billets and bars, such as lime, oxygen, nitrogen, coal dust, oil, gas and refractory materials. A deeper analysis could generate additional interesting information. Furthermore, transport related emissions could be higher when going beyond a regional scale analysis. In fact, if the complete paths from the origin of raw materials to the plant were taken into account, the produced emissions would probably be higher than those assessed.

As a consequence, solutions for lowering the emissions from transport related to the industrial activity like the one analyzed should be promoted.

PM emissions from transport seems to play a secondary role in case of incineration when compared with stack emissions

REFERENCES

1.   B. Brunekreef, S. Holgate, "Air pollution and health", in The Lancet, vol.360/9341, 2002, pp. 1233-1242.

2. European Community, "Council Directive 1999/39/EC of 22 April 1999 regulating to limit values for sulphur dioxide, nitrogen dioxide and oxides of nitrogen, particulate matter and lead in ambient air", in Official Journal of the European Communities, L163, 1999, pp. 41–60.
3. European Community, "Council Directive 2008/50/EC, on ambient air quality and cleaner air for Europe". In Official Journal of the European Communities, L152, 2008, 1.
4. D. Heimann, M. de Franceschi, S. Emeis, P. Lercher, P. Seibert, 2007. "Air Pollution, Traffic Noise and Related Health Effects in the Alpine Space − A Guide for Authorities and Consulters. ALPNAP comprehensive report", Università degli Studi di Trento, Dipartimento di Ingegneria Civile e Ambientale, Trento, Italy, pp. 335.
5. ISTAT, 2011. http://demo.istat.it/bil2008/index.html
6. F. Barbone, F. Brevi, U. Ghezzi, M. Ragazzi, A. Ventura, 2009. "Concessione di lavori per la progettazione, realizzazione e gestione dell'impianto di combustione o altro trattamento termico con recupero energetico per rifiuti urbani e speciali assimilabili in località Ischia Podetti, nel Comune di Trento − Studio di Fattibilità", Provincia Autonoma di Trento, Comune di Trento, Italy, pp. 226.
7. European Environmental Agency, "EMEP/CORINAIR Emission Inventory Guidebook 2007", Technical Report, no. 16, 2007, Bruxelles.

CHAPTER 3

A Critical Analysis of Emissions and Atmospheric Dispersion of Pollutants from Plants for the Treatment of Residual Municipal Solid Waste

GABRIELA IONESCU, DINO ZARDI, WERNER TIRLER, ELENA CRISTINA RADA, AND MARCO RAGAZZI

3.1 INTRODUCTION

The release of substances that affect air quality is to be included in the consequences of the increase of waste production, typical for our consumer society. A part of these substances is classified as air pollutants and thus noxious for the environment and human health [1]. Another part of them is classified as bad-smelling and so responsible of olfactory nuisance [2]. Last but not least, other substances are greenhouse gases, producing modifications in the radiation budgets controlling the earth surface temperature [3]. Assessing air pollution in complex morphologies becomes an important issue in order to implement mitigation measures and limit

A Critical Analysis of Emissions and Atmospheric Dispersion of Pollutants from Plants for the Treatment of Residual Municipal Solid Waste. Ionescu G, Zardi D, Tirler W, Rada EC, and Ragazzi M. UPB Scientific Bulletin, Series D: Mechanical Engineering *74,4 (2012). http://scientificbulletin.upb.ro/ rev_docs_arhiva/rez4b0_198021.pdf. Reprinted with permission from the authors and the publishers.*

emissions [4]. Various research investigations conducted up to date have demonstrated that with respect to the self-ignition tendency i.e. tendency towards low temperature oxidation, solid waste represents a very complex and sensitive material [5]. The evolution of the regulations in the European Union, in the field of municipal solid waste (MSW) management, has improved the overall scenario but some aspects must be developed.

The aim of this study is to evaluate some critical aspects related to the role of MSW composition in air emission, the release of some air pollutants, the processes governing the dispersion of pollutants into the atmosphere and a case study of a RMSW incinerator.

3.2 ROLE OF MSW COMPOSITION IN AIR EMISSION

The composition of the incinerated feedstock stream has a direct influence on the Waste-to-Energy (WtE) technological conversion process and its environmental impact [2,6]. Primarily, in any waste thermal treatment, two aspects must be taken into account: the overall composition of MSW and the composition of the residual MSW (RMSW). Differences are related to the role of selective collection. The latter will influence the waste incineration as [7]:

- some materials should not be incinerated (they are more suitable for recycling, they are non-combustible or their by-products may be harmful);
- poor operating practices and the presence of chlorine in the MSW may lead to emissions containing highly toxic dioxins and furans (PCDD/F);
- the control of metal emissions may be difficult for inorganic wastes containing heavy metals.

This paper focuses on emissions from RMSW but takes into account also the trend in the composition of MSW in an European region where waste management is enhanced thanks to the adoption of the EU criteria: the North-East of Italy. In Table 1 some considerations on the trend in the MSW composition in the selected region are reported.

A part of these considerations can be applied also to other Italian regions and other European areas where the economic development is similar and the MSW management is close to be optimum according to the EU principles.

TABLE 1: Trend of MSW composition in the North-East of Italy

MSW Fraction	MSW composition trend
Food waste	The per-capita generation is nearly steady as the nutritional behaviour does not changes, significantly among the population. Its percentage has decreased in the last years as other MSW fractions increased (in particular, packaging showed a significant increase in the last decades). The evolution of the tourism sector in some areas can give an additional increase of food waste generation.
Garden waste	In areas where domestic composting has been adopted, the amount of garden waste collected has decreased slightly. This option has been favored through the decrease of the tariff to be paid for domestic waste management.
Paper and cardboard	The recent economical crisis stopped the increase of cellulosic packaging that characterised the recent decades. A few initiatives for reducing waste generation are expected to slightly decrease its percentage in MSW.
Plastics	Also in this case the recent economical crisis stopped the increase of plastic packaging. A few initiatives for reducing waste generation are expected to slightly decrease its percentage in MSW. In particular, the substitution of plastic shoppers with reusable bags is a strategy recently introduced on a large scale.
Glass	Plastic bottles are more and more preferred to glass bottles because of their weight. Glass percentage in MSW decreased in the last decades.
Metals	In the last decades the use of Aluminium cans increased.
Exhausted batteries	An important limitation to the production of Hg batteries was introduced, with consequences on its presence in MSW. The introduction of rechargeable batteries is giving additional results in terms of content decrease.
Composites	Composite materials have been more and more introduced as packaging, with some troubles for their recycling viability.
Other fractions	The other fractions present are not relevant in terms of trend.

As a consequence of the explained characteristics, a general trend in Lower Heating Value (LHV) increase can be observed: its present value is around double compared to the one in the 70s.

In Table 2 the analysis concerns RMSW, obviously affected by the strategies of selective collection (SC). To this concern the SC in the North-East of Italy has reached very high values. In many municipalities the RMSW has been reduced to one third (or less) of the generated MSW. The present priority is to improve the quality of selection at the source. In particular the stream of light packaging is affected by significant mistakes of separation.

TABLE 2: Trend of RMSW composition in the North-East of Italy

RMSW	RMSW composition trend
Food waste	The percentage of food waste in the RMSW is quickly decreasing because of the implementation of selective collection of the organic fraction. In some areas the rate of collection at the source is expected to reach 80% of this fraction. The selective collection of other fractions partially counter-balances this phenomenon. The resulting percentage of food waste can decrease to values around 10% in some extreme cases. A consequence is the decrease of Cl (from NaCl) in RMSW.
Garden waste	Where food waste selective collection is implemented also garden waste is generally segregated, thus the considerations are similar to the ones above (Cl excluded).
Paper and card-board	The implementation of selective collection has decreased their percentage in the RMSW. A 100% interception is not reachable as a part of paper is not suitable for recycling because of the presence of pollutants or undesired moisture.
Plastics	Only plastic packaging is selectively collected, thus the presence of plastics in RMSW is still significant. This aspect contributes to a high LHV of the RMSW, even higher than the limit of 13MJ/kg set from EU for classifying a waste non-suitable for landfilling.
Glass	Selective collection of glass has reached very high values thanks to the citizen behaviour used to keep glass separated from other fractions for safety reasons.
Metals	Selective collection gives low % of metals in RMSW.
Exhausted batteries	The decrease in Hg content has important consequences on Hg emissions in case of RMSW burning.
Diapers	The decrease of the RMSW amount (even down to 35% of the generated MSW) can point out the presence of non-recyclable diapers (in percentage).
Composites	Only in some areas a few composites can be recycled; thus their presence in RMSW varies significantly.
Other hazardous waste	A good management of municipal hazardous waste, through source separation, gives a lower presence of contaminants in the RMSW. Expired medicaments remaining in the RMSW are close to zero.
Other fractions	The other fractions present are not relevant in terms of trend.

Summing up, the described trend has a few consequences on the characteristics of RMSW: lower moisture; lower putrescibility; lower Hg and other persistent pollutants; lower Cl content; higher LHV. Concerning LHV, the increase of its value does not mean always an increase of energy available for a WtE plant, as selective collection diverts also fractions with a high energy content (such as plastics and paper).

3.3 CRITICAL ANALYSIS OF AIR POLLUTANTS RELEASE BY RMSW MANAGEMENT

The atmospheric emission factors from RMSW treatment or disposal depend on: (a) presence of pollutants in the waste [8]; (b) prevention of pollutant generation [9]; (c) pollutant removal/ conversion after generation [10]. In order to develop a critical analysis of pollutants released into the atmosphere from RMSW options, the following steps were adopted:

- Step 1. Seeking of available options for RMSW treatment.
- Step 2. Selection of options suitable for real scale plants. After this selection the following options remained:
    - biostabilization [11]
    - bio-drying [12]
    - anaerobic digestion [13]
    - combustion [2]
    - gasification [10]
    - pyrolysis [10]
    - landfilling [9]

These options must be seen as a part of an integrated system [14][15] [16] but in this section the characteristics of each option are analyzed independently on the presence of other processes.

- Step 3. Seeking of literature data on emission factors to air from RMSW treatment (only for the above selected options), avoiding commercial data. Only data from ISI/Scopus articles were selected in order to have validated information.
- Step 4. Re-organization of data in three categories:
    - non carcinogenic pollutants involving local impacts (these
    - pollutants can be relevant both for hourly peaks and for yearly averages)
    - carcinogenic pollutants involving local impacts (these pollutants are relevant only for long time human exposure as yearly averages)
    - global warming pollutants (these pollutants, named greenhouse gases, are not strictly relevant locally as they involve global environmental processes)
- Step 5. Critical analysis of data previously selected in order to point out the trend of the RMSW sector. As a result, the Table 3 has been generated.

TABLE 3: Main critical aspect related to emission factors (EFs) to air from RMSW processes

	Non-carcinogenic pollutants (local)	Carcinogenic pollutants of local interest	Global warming pollutants
Biostabilization (biological treatment before landfilling) [17]	Odour emissions can be critical during the maturation stage if the accelerated fermentation is not Optimized. NOx from Rigenerative Thermal Oxidation (RTO) could be relevant, if adopted, because of methane burning; EFs can be even in the order of hundreds of mg_{NOx}/kg_{waste}.	PCDD/F already present in RMSW could be partially stripped and released. This phenomenon, recently pointed out, is not yet taken into account in all the EU countries. PCDDF EFs could be in the order of tens of pg_{TEQ}/kg_{waste}. The variability depends on the contamination of waste and on the air treatment line.	N_2O can come from biological processes; as a consequence, in spite of the biological approach, greenhouse gas emissions are not zero (an additional contribution comes from electricity consumption)
Biodrying (biological pre-treatment for energy options) [3,18, 19]	NOx from Rigenerative Thermal Oxidation could be relevant, if adopted. EFs can be even in the order of hundreds of mg_{NOx}/kg_{waste}.	PCDD/F already present in RMSW could be partially stripped. Even if lower than biostabilization, PCDDF EFs could be in the order of tens of pg_{TEQ}/kg_{waste}.	Similar to biostabilization but with lower emissions (shorter process)
Anaerobic digestion (biological treatment generating methane) [13,20]	NOx emissions from biogas combustion can be not negligible.	VOCs from biogas burning should be minimized; PCDD/F could be also generated. PCDDF EFs could be in the order of tens of pg_{TEQ}/kg_{waste}.	Not critical, if compared to other biological processes (CO_2 is biogenic)
Combustion (thermochemical oxidation) [21,22,23]	NOx EFs could be relevant if specific removal stages are not adopted. EFs can be even in the order of g_{NOx}/kg_{waste} in unoptimised plants, but < 300 mg_{NOx}/kg_{waste} in the best plants.	PCDD/F and Cd are the most relevant compounds. PCDDF EFs could be in the order of hundreds of pg_{TEQ}/kg_{waste} if the plant operates close to the allowed limits. Some plants can decrease EFs of one order of magnitude.	Non-biogenic CO_2 EFs depend on RMSW
Gasification (process aimed to partial oxidation) [24]	CO emissions, after syngas exploitation, can be not negligible.	Heavy metals EFs could be high if high temperature is adopted	Non-biogenic CO_2 emissions depend on RMSW composition
Pyrolysis (endothermal process) [25]	Char combustion can generate high emissions for many pollutants in plants not optimized for that	Stripping of PCDD/F already present in RMSW could give unexpected EFs (even in the order of ng_{I-TEQ}/kg_{waste})	Net CO_2 EFs depend on RMSW composition and on the overall strategy for the pyrolysis outputs
Landfilling (putrescible waste no longer allowed) [26,9]	Odour problems if not correctly managed (residual putrescibility)	PCDD/F emissions from bad quality biogas burning in small landfills	CH_4 fugitive emissions if not correctly managed (residual putrescib.)

3.4 PROCESSES GOVERNING THE DISPERSION POLLUTANTS INTO THE ATMOSPHERE

The local interest is related to human exposure. Its level depends on the climatology of the area where a plant is located, on the amount of pollutants released into the atmosphere and on the way they are released [17,27]. The following sections of this paper deal with these concepts.

The fate of pollutants released into the atmosphere, through waste handling, treatment and storage processes, are determined by various combinations of weather and climate factors. Air contaminated with pollutants released into the atmosphere rises up to the level of neutral buoyancy. There, it is advected by the horizontal mean wind components, which usually exhibit an increasing speed at higher levels (the so called wind shear). Whole being advected, contaminants get mixed by smaller eddies generated by turbulence motions. The combined effect of advection by the mean flow and diffusion by continuous and chaotic turbulence movements determines the so called dispersion.

Turbulence can be generated either mechanically, i.e. by the interaction between the sheared wind and turbulent eddies, or thermally, i.e. through the buoyancy effect induced by the warming of cooling of the air close to the ground, mainly as a consequence of the surface heat budget following the diurnal cycle.

The physical processes explained above are mathematically described by the equation governing the local evolution in time t of the mean concentration c (i.e. averaged with respect to the turbulent fluctuations c'). The rate of change of c (eq. (1)) is determined by three factors, i. e. respectively (a) the advection by the mean wind, represented as the scalar product of the wind velocity u by the gradient of the mean concentration, (b) the divergence of the turbulent flux u'c' determined by the coupling between turbulent fluctuations of wind velocity and concentration, and (c) the sources (e.g. photochemical production) or sinks (e.g. removal by rain, snow or deposition on receptors) of air pollutants.

$$\frac{\partial \bar{c}}{\partial t} = \underbrace{-\bar{u} \cdot \nabla \bar{c}}_{advection} \quad \underbrace{-\nabla \cdot \overline{u'c'}}_{turbulent\ diffusion} \quad \underbrace{+S}_{production/removal} \tag{1}$$

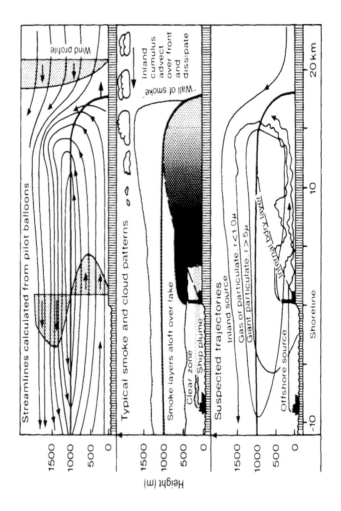

FIGURE 1: The form of the lake Breez front near Chicago measured using tetroons and aircraft. Special features are the "wall of smoke" and the evidence for recirculation [28]

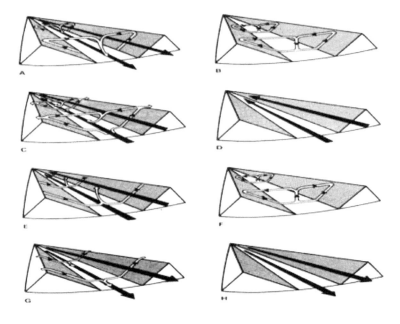

FIGURE 2: Diurnal cycle of valley winds: (a) Sunrise: onset of upslope winds (white arrows); continuation of mountain wind (black arrows). Valley is cold, plains are warm. (b) Forenoon (about 0900): strong slope winds, transition from mountain wind to valley wind. Valley temperature is same as plain. (c) Noon and early afternoon: diminishing slope winds; fully developed valley wind. Valley is warmer than plains. (d) Late afternoon: slope winds have ceased, valley wind continues. Valley is still warmer than plain. (e) Evening: onset of down-slope winds, diminishing valley wind. Valley is slightly warmer than plains. (f) Early night: well developed down-slope winds, transition from valley wind to mountain wind. Valley and plains are at the same temperature. (g) Middle of the night: down-slope winds continue, mountain wind fully developed. Valley is colder than plains. (h) Late night to morning: down-slope winds have ceased, mountain wind fills valley. Valley is colder than plain [28].

It is clear that the wind structure is a key factor in determining the fate of pollutants. However this structure is determined by the scale at which air motions are generated. Usually atmospheric motions are hierarchically categorised into four main scales. The largest one is known as planetary scale, as it includes motions occurring on an extent embracing all or most of the planet, as part of the so called general circulation of the atmosphere). A smaller scale is determined by typical features of the pressure and temperature field, such as high and low pressure structures (anticyclones and cyclones respectively) often associated with moving fronts, i.e. regions of sharp temperature contrasts between adjacent air masses displaying different mean temperatures. This scale is a synoptic scale. A typical horizontal length for this scale is of order of some thousands kilometres. The so called mesoscale includes phenomena occurring in a range of some tenths to some hundreds kilometres. It includes phenomena determined by geography (mountains, deserts, sea and lake shore). Typical phenomena falling in these categories are sea breezes and diurnal mountain and valley winds detailed in Fig. 1 and 2. Then there is the broad range of local scale phenomena, including urban effects, roughness-induced phenomena, turbulent motions.

In the presented case study, the Italian territory is covered by a 23.2% of plain areas, 41.5% by hills and 35.2% by mountains. About half of the population lives in plain regions (49.7%), but the remaining half live hills in (37.3%) or mountains 13%). Furthermore, it is bounded by 8490 km of maritime coasts. The local climatology can strongly affect the dilution of pollutants emitted from a plant treating RMSW. The selection of a site for a new plant can play an important role: from site to site the local impact can vary of one order of magnitude.

3.5 CASE STUDY

A typical case useful for understanding the local impact of the emissions from a RMSW treatment plants is the one concerning an incinerator. This impact depends on many factors: site characteristics (in particular climatology); amount of RMSW to be treated; LHV of the RMSW; emission concentrations at the stack during operating conditions; stack height; off-gas velocity at the stack; off-gas temperature at the stack; yearly operating hours.

Thanks to the optimization of all these parameters, a modern incinerator can have a negligible impact on the interested area. An example of that can come from the case of the incinerator of Bolzano. In this paper, a zoom on the role of the most relevant carcinogenic pollutants is presented. In Fig. 3 some results of the modelling of pollutant dispersion and deposition are presented. The present results come from an integrated approach developed at the Civil and Environmental Department of the University of Trento.

These results suggest that the process of transport and dispersion of pollutants from the operation of the incineration plant crucially depends on the diurnal cycle of the valley winds. In the case of Northerly wind, the process is dominated by the flow channel along the Adige valley where the maximum concentration values at ground level occur south of the incinerator at a distance of about 3 km. The scenario becomes more complicated in the case of wind flow from the southern sector: as stated, the main flow is directed, depending on the atmosphere conditions that exist in the valley town, towards the valley of Isarco or to Merano. The average year shows that this phenomenon gives rise to two areas of maximum concentration that take place near the town of Bolzano, the first North-West part of the town and the second located approximately in the centre of the town.

In particular, Fig. 3 concerns the contribution of Cd emissions to the air concentration at ground level. For this parameter, the deposition [29] is less relevant as its impact is most significant in case of inhalation. The highest contribution is only about $1.1 \cdot 10^{-4}$ ng/m^3 whilst in EU its limit in air is 5 ng/m^3.

Concerning the contribution of PCDD/F to the concentration in air at ground level, the highest values are in urbanized areas but its contribution is so low that this impact has been considered acceptable: the peak value resulted only 0.04 fgTEQ/m^3. A verification of this amount was made measuring the PCDD/F air and soil concentrations in the area affected by the plant. The measured value in air resulted 67 fgTEQ/m^3 demonstrating that a modern incinerator can give a negligible contribution to the presence of PCDD/F in an area.

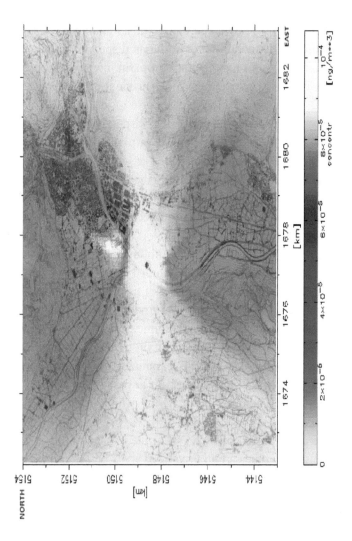

FIGURE 3: Contribution to the Cd average yearly concentration in air at ground level

Moreover, additional directional (wind dependent) samplings in the following years confirmed that PCDD/F in air is not critical in the Bolzano area and there is not a significant influence from the incinerator.

A confirmation of the low impact was made also characterizing the soil in the area; the obtained value was 1.23 ng_{TEQ}/kg, lower than the limit for residential areas (10 ng_{TEQ}/kg). Additional samplings in the urban area confirmed that the typical values are generally lower than 2 ng_{TEQ}/kg.

Taking into account only the contribution of PCDD/F to the air concentration, an underestimation of the impact of these pollutants can emerge. Indeed an important aspect concerns the deposition of PCDD/F and their role depending on the local food production and consumption. In the case study the deposition results gave a peak value as 13 $pgTEQ$ m^{-2} y^{-1}. A guideline value come from Flanders [30], with a value of 1,241 pg m^{-2} y^{-1} demonstrating the low impact of the plant.

As a carcinogenic pollutant has never a health risk equal to zero, it is important to assess the incremental risk related to a cancer event referred to the most exposed person in the area and the overall risk. The results of the adoption of a health risk method demonstrated that the individual one is $<< 10^{-6}$ and the overall one is $<<1$.

The positive results of the above impact analysis allowed the local administration to propose a new RMSW combustion plant in the same site in order to substitute the existing one, as approaching its end of life. As pointed out in Table 3, the NOx emission minimization was assumed as a priority for the new plant. The incinerator under construction will guarantee average daily NOx concentration at the stack lower than 40 mg/Nm^3, significantly lower than the limit requested at national level.

These positive results are related to an optimized plant. In the sector of waste management sub-optimum design solutions can give not negligible local impact also when incineration is not adapted [6]. Both for bio-mechanical plants and for landfilling, the release of pollutants at ground level can suffer from a bad dilution that does not protect the residential areas in the surroundings. A demonstration of the not optimized way of release from a biofilter comes from the TOC parameter: the German regulation on bio-mechanical plants [6] sets a TOC emission factor value as a limit to be complied with. For the TOC parameter it is asked to emit less than 55 mg/kg_{RMSW}. If we take into account that an incinerator could be authorised to have at the stack 5 mg/Nm^3, in the case of a specific air flow-rate of 6 Nm^3/kg$_{RMSW}$ we could have an emission factor to air of 5 x 6 = 30 mg/kg$_{MSW}$, lower than the one for bio-mechanical plants (30 < 55 mg/kg$_{RMSW}$). It is clear that, with similar emission factors but different height of release, the local impact of the incinerator will be significantly lower, for this parameter.

From these considerations, a modification of the regulations concerning low level emissions from MSW treatment plants is compulsory. To this concern, it must be pointed out that a significant number of composting plant have been recently stopped in Italy because of the unoptimized release of process air. In this case the odour parameter gives a direct idea of the bad managed impact of the plant.

3.6 CONCLUSIONS

The contents of this paper demonstrate that trends of the environmental impact in the MSW sector can be found thanks to the evolution of the waste management strategies. However some prevention activities in terms of emissions to the atmosphere cannot decrease the importance of the analysis of the local meteorology in order to verify the compatibility of a site with the proposal of a plant. A problem to be faced with can be the lack of meteorological data if the region of the site has not an adequate tradition of meteorological data collection. In this case a specific campaign of at least one year should be planned. Moreover, in the MSW sector there is a unbalanced approach from the regulatory point of view, as the role of low level emissions is not managed with the same attention of the case of emissions from a high stack. In this last case, dispersion and deposition modelling demonstrate that modern incinerators may be constructed keeping very low their local impact. This result cannot be generalized as the role of meteorology can significantly change the local incidence of a plant in terms of air pollution. Additionally the concept of "modern" must be related to low emission values really guaranteed at the stack.

REFERENCES

1. D.M. Cocarta, E.C. Rada, M. Ragazzi, A. Badea and T. Apostol, "A contribution for a correct vision of health impact from municipal solid waste treatments", Environ. Technol., vol. 30, no. 9, 2009, pp. 963-968.
2. E.C. Rada, I.A. Istrate and M. Ragazzi, "Trends in the management of residual municipal solid waste", Environ. Technol., vol. 30, no. 7, 2009, pp. 651-661.
3. M. Ragazzi, E.C. Rada and D. Antolini, "Material and energy recovery in integrated waste management systems: An innovative approach for the characterization of the gaseous emissions from residual MSW bio-drying", Waste Manage. vol. 31, no. 9-10, 2011, pp. 2085-2091.
4. S. Ciuta, M. Schiavon, A. Chistè, M. Ragazzi, E.C. Rada, M. Tubino, A. Badea and T. Apostol, "Role of feedstock transport in the balance of primary PM emissions in two case-studies: RMSW incineration vs. Sintering plant", U.P.B. Sci. Bull., serie D, vol. 74, no. 1, 2012, pp. 211-218.
5. A. Jovovic, G. Vujic, M. Pavlovic, D. Radic, D. Jevtic and M. Stanojevic, "Spontaneous Ignition/Low Temperature Oxidation of Municipal Solid Waste", Rev. Chim. (Bucharest), vol. 62, 2011, pp.108-112
6. E.C. Rada, M. Ragazzi, V. Panaitescu and T. Apostol, "Some research perspectives on emissions from bio-mechanical treatments of municipal solid waste in Europe", Environ. Technol., vol. 26, no. 11, 2005, pp. 1297-1302
7. G. McKay, " Dioxin characterisation,formation and minimisation during municipal solid waste (MSW) incineration: review", Chem Eng J., vol. 86, 2002, pp. 343-368.
8. E.C. Rada, A. Franzinelli, M. Ragazzi, V. Panaitescu, T. Apostol, "Modelling of PCDD/F release from MSW bio-drying", Chemosphere, vol. 68, no. 9, 2007, pp. 1669-1674.
9. D.P. Komilis, R.K. Ham, R.K. and R. Stegmann, "The effect of municipal solid waste pretreatment on landfill behavior: a literature review", Waste Manage Res., vol. 17, no. 1, 1999, pp 10-19.
10. M. Ragazzi, E.C. Rada, "Multi-step approach for comparing the local air pollution contributions of conventional and innovative MSW thermo-chemical treatments", Chemosphere, vol. 89, no.6, 2012, pp. 694-701
11. E.C. Rada, M. Ragazzi,V. Panaitescu,T. Apostol," The role of bio-mechanical treatments of waste in the dioxin emission inventories, Chemosphere, vol. 62, no. 3, 2006, pp. 404-410
12. E.C. Rada, M. Ragazzi, A. Badea,"MSW Bio-drying: Design criteria from A 10 years research", U.P.B. Sci. Bull., serie D, vol. 74 , no. 3, 2012, pp. 209-216
13. E.C. Rada and M. Ragazzi, "Critical analysis of PCDD/F emissions from anaerobic digestion", Water Sci Technol., vol. 58, no. 9, 2008, pp. 1721-1725.
14. E.C. Rada, M. Venturi, M. Ragazzi, T. Apostol, C. Stan, C, Marculescu, " Bio-drying role in changeable scenarios of Romanian MSW management", Waste and Biomass Valor., vol. 1, no. 2, 2010, pp. 271-279.

15. E.C. Rada, I.A. Istrate,V. Panaitescu, M. Ragazzi, T.M. Cirlioru, T. Apostol, "A comparison between different scenarios of Romanian municipal solid waste treatment before landfilling", Environ. Eng. Manage. J., vol. 9, n. 4, 2010, pp. 589-596.

16. E.C. Rada, M. Ragazzi, V. Panaitescu, "MSW bio-drying: An alternative way for energy recovery optimization and landfilling minimization", U.P.B. Sci Bull., serie D., vol. 71, no. 4, 2009 , pp. 113-120.

17. Q. Liu, M. Li, R. Chen, Z.Y. Li, G.R. Qian, T.C. An, J.M. Fu and G.Y. Sheng, "Biofiltration treatment of odors from municipal solid waste treatment plants ",Waste Manage., vol. 29, no. 7, 2009, pp. 2051-2058

18. E.C. Rada, M. Ragazzi, D. Zardi, L. Laiti and A. Ferrari, " PCDD/F enviromental impact from municipal solid waste bio-drying plant", Chemosphere, vol. 84, no. 3, 2011, pp. 289-295.

19. C.A. Velis, P.J. Longhurst, G.H. Drew, R. Smith and S.J.T. Pollard, "Biodrying for mechanical-biological treatment of wastes: A review of process science and engineering "Bioresour. Technol., vol. 100, no. 11, 2009, pp. 2747-2761.

20. M.A. Latif, A. Ahmad, R. Ghufran and Z.A. Wahid, "Effect of temperature and organic loading rate on upflow anaerobic sludge blanket reactor and CH4 production by treating liquidized food waste", Environ. Progr. Sustain. Energy, vol. 31, no. 1, 2012, pp. 114-121.

21. S. Consonni and F. Vigano, "Waste gasification vs. conventional Waste-To-Energy: A comparative evaluation of two commercial technologies", Waste Manage., vol 32, no. 4, 2012, pp. 653-666.

22. C. Marculescu, G. Antonini and A. Badea, "Analysis on the MSW thermal degradation processes", Global Nest J., vol 9, 2007, pp. 57-62.

23. M.J. Quina, J.C. Bordado and R.M. Quinta-Ferreira, "Treatment and use of air pollution control residues from MSW incineration: An overview", Waste Manage., vol. 28, 2008, pp. 2097-2121.

24. U. Arena, "Process and technological aspects of municipal solid waste gasification. A review", Waste Manage., vol. 32, no. 4, 2012, pp. 625-639.

25. P. Baggio, M. Baratieri, A. Gasparella and G.A. Longo, "Energy and environmental analysis of an innovative system based on municipal solid waste (MSW) pyrolysis and combined cycle", Appl. Therm. Eng., vol. 28, no. 2-3, 2008, pp. 136-144.

26. D.B. Yue, J.G. Liu, P. Lu, Y. Wang, J. Liu and Y.F., "Release of non-methane organic compounds during simulated landfilling of aerobically pretreated municipal solid waste", J Environ Manage., vol. 101, 2012, pp. 54-58.

27. L. Giovannini, D. Zardi and M. Franceschi, "Analysis of the Urban Thermal Fingerprint of the City of Trento in the Alps", J. Appl. Meteorol Climatology, vol. 50, no. 5, 2011, pp. 1145-1162.

28. J.E. Simpson, "Sea breeze and local winds", Water Sci Technol., vol. 58, 2008, pp. 1721-1725.

29. A. Lucaciu, C. Motoc, M. Jelea and S.G. Jelea,. "Survey of heavy metal deposition in Romania: Transylvanian plateau and western carpathians mountains", U.P.B. Sci. Bull., vol. 72, no. 2, 2010, pp. 171-178.

30. L. Van Lieshout, M. Desmedt, E. Roekens, R. De Fré, R. Van Cleuvenbergen and M. Wevers, "Deposition of dioxins in Flanders (Belgium) and a proposition for guide values", Atmos. Environ., vol. 35, 2001, pp. 83-90.

PART III

VEHICLE AND
TRANSPORTATION EMISSIONS

CHAPTER 4

Performance of a Diesel Engine with Blends of Biodiesel (from a Mixture of Oils) and High-Speed Diesel

HIFJUR RAHEMAN, PRAKASH C. JENA, AND SNEHAL S. JADAV

4.1 BACKGROUND

Higher soot or carbon deposits on in-cylinder engine components and lubricating oil contamination are the main causes for engine wear. Wear processes due to oil contamination lead to diminished fuel efficiency, shorter useful oil service life, reduced component life, and loss of engine performance. Hence, in addition to engine performance and emissions of diesel engine, soot depositions on engine components and wear metal additions in lubricating oil of the engine are required for the selection of a fuel that would replace conventional diesel fuel. With increase in demand for using biodiesel in place of high-speed diesel (HSD), a detailed study on these as-

Performance of a Diesel Engine with Blends of Biodiesel (from a Mixture of Oils) and High-Speed Diesel. © *Raheman H, Prakash C Jena PC, and Jadav SS.* International Journal of Energy and Environmental Engineering *4,6 (2014), doi:10.1186/2251-6832-4-6. Licensed under a Creative Commons Attribution 2.0 Generic License, http://creativecommons.org/licenses/by/2.0/.*

pects is very much required for the selection of a suitable blend to replace HSD for long-term use in diesel engines.

Very few studies have been conducted on soot depositions and wear metal additions in the lubricating oil of engine that are operated with biodiesel obtained from single-vegetable oil and its blends with HSD. It was reported that metal addition in the lubricating oil of engines operated with different biodiesel (soybean, palm kernel oil, linseed oil) blends was lower or similar as compared to when the engine was operated with HSD alone [1–5]. Ramaprabhu et al. [6] reported that iron wear was almost similar, whereas copper wear was higher in the lubricating oil of diesel engine operated with Jatropha and Karanja biodiesel blend as compared with when the lubricating oil of diesel engine was operated with HSD. Lesser soot deposition on engine components was reported for biodiesel blend-operated diesel engines [6, 7]. With increasing demand on the use of biodiesel, more and more oils and mixtures of oils are explored for biodiesel production [8]. Mahua (*Madhuca indica*) oil (MO) with high free fatty acids (FFA) and simarouba (*Simarouba glauca*) oil (SRO) are few such potential oils suitable for biodiesel production. Higher FFA present in oil requires higher methanol; hence, an attempt was made to produce biodiesel from a mixture (50:50) of these two oils to reduce methanol requirement in biodiesel production [9]. Performance evaluation, emission examination along with soot deposition on in-cylinder engine components, and wear metal addition in the lubricating oil of the engine are required for biodiesel obtained from such mixture of oils. Hence, a study was undertaken at Indian Institute of Technology, Kharagpur, India.

4.2 METHODS

4.2.1 BIODIESEL PRODUCTION

MO and SRO with FFA levels of 13% and 1.43%, respectively, were mixed at 50:50 v/v proportions (MSO) to reduce methanol consumption in biodiesel production. This oil mixture (MSO) had an FFA content of 7.19%. Hence, pretreatment esterification with an acid catalyst to bring down the FFA level of the oil mixture to around 1% was required.

TABLE 1: Technical specification of engine and hydraulic dynamometer

	Particulars	Details
Engine	Model	DM14
	Maximum power (kW)	10.3
	Type	Water-cooled, four stroke
	Rated speed (rpm)	1,500
	Number of cylinders	1
	Compression ratio	15.5:1
	Bore × stroke (mm)	114.3 × 116
	Brake mean effective pressure at 1,500 rpm (kg/cm²)	7.054
	Combustion	Direct injection (DI) and naturally aspirated
	Injection timing	24° before TDC
Hydraulic dyna-mometer	Model	AWM15
	Type	Hydraulic
	Water pressure at inlet (kg/cm²)	1.5
	Power (hp)	
	Range	1 to 100
	Accuracy	±2%
	Data resolution	0.02%
	Maximum speed (rpm)	
	Range	5,650 to 8,000
	Accuracy	±2%
	Data resolution	0.03
Exhaust gas analyzer	Model	PEA205
	CO (%)	
	Range	0 to 15
	Accuracy	±0.06
	Data resolution	0.001
	HC (ppm)	
	Range	0 to 30,000
	Accuracy	±4%
	Data resolution	1
	NOx (ppm)	
	Range	0 to 5,000
	Accuracy	±2%
	Data resolution	1

A two-step 'acid-base' process, an acid pretreatment followed by the main base-transesterification reaction, using methanol as reagent and H_2SO_4 and KOH as catalysts for acid and base reactions, respectively, was followed to produce biodiesel from MSO. The various fuel properties of MO, SRO, MSO, and biodiesel (B100) obtained from MSO and its blends with HSD (B10, B20) were determined as per the American Society for Testing and Materials (ASTM) standards.

4.2.2 EXPERIMENTAL SETUP

Performance and exhaust emissions were studied for a 10.3-kW single-cylinder four-stroke direct-injection water-cooled diesel engine using bio-diesel blends with HSD (B10 and B20) by varying the engine load from no-load to 100% load in steps of 20%. A hydraulic dynamometer (SAJ Model AWM 15, SAJ International Pvt. Ltd, Pune, India; 100 hp at 5,650 to 8,000 rpm) equipped with a strain gauge-based load cell and digital readout for measuring engine torque and speed was used for loading the diesel engine. Details of the technical specification for the engine, dynamometer, and exhaust gas analyzer used are given in Table 1.

4.3 METHODOLOGY

4.3.1 PERFORMANCE TESTS AND EMISSION MEASUREMENT OF DIESEL ENGINE WITH BIODIESEL BLENDS

Performance tests were conducted in a diesel engine using HSD and blends of biodiesel (B10 and B20) following the Indian standard (IS) 10000: part 8 [10]. The rated engine load measured was found to be 57.8 Nm at 1,500 ± 10 rpm, and the corresponding rated power developed by the engine was 9.1 kW. At rated power, the load going to the engine was taken as 100%. Accordingly, intermediate loads were calculated as 11.56, 23.12, 34.68, and 46.24 Nm (corresponding to 20%, 40%, 60%, and 80% of torque obtained at rated power). The tests were conducted for 3 h and 30 min for each test fuel, and load was applied at an interval of 30 min. The readings were recorded at an interval of 10 min for a particular engine load. During this test, performance parameters such as brake specific fuel consumption (BSFC), brake thermal efficiency (BTE), and exhaust gas temperature (EGT) were determined by measuring fuel consumption, engine torque, and EGT. During this performance test, exhaust emissions such as CO, hydrocarbon (HC), and oxides of nitrogen (NOx) were also measured using an online exhaust gas analyzer. A suitable blend for long-term use was selected by comparing the performance and emissions of the two blends with those of HSD.

4.3.2 ESTIMATION OF SOOT DEPOSITS ON IN-CYLINDER ENGINE COMPONENTS

The engine was run for 100 h (16 test cycles each of 6.25-h continuous running) with each of the two fuels (HSD and selected biodiesel blend). In 100 h, the engine was subjected to different loadings, and the lubricating oil samples were collected at an interval of 25 h of engine run. The duration for each loading was decided as per IS 10000: part 9 [11]. In each test cycle, loads were applied in a random manner for predetermined durations, i.e., 100%, 50%, 100%, no load, 100%, and 50% load for 93.75 (including 11.72 min of warm up), 93.75, 23.45, 11.72, 70.31, and 82 min, respectively. After 100 h of engine operation with each of these fuels, the engine was dismantled, and the cylinder head, piston, and fuel injector were removed carefully and kept on a clean surface. Photographs of these components were taken to visually compare the soot deposits when operated with different fuels. Later, the soot deposits were gently scraped from these components using a wooden scraper. Weights of these scraped deposits for each component were taken separately for comparison.

4.3.3 DETECTION OF WEAR METALS BY ATOMIC ABSORPTION SPECTROSCOPY ANALYSIS

The lubricating oil samples collected after each 25 h of engine run were subjected to atomic absorption spectroscopy (AAS) analysis to determine the addition of metals (such as copper, zinc, iron, manganese, nickel, lead, magnesium, and aluminum) in lubricating oil of engine. Dry ashing technique was used to prepare the sample of lubricating oil for metal analysis by AAS. Each lubricating oil sample was taken in a 250-ml conical flask and thoroughly mixed in a water bath at 50°C for 1 h at constant speed. Approximately 10 g of mixed lubricating oil sample was taken in a previously washed and dried silica crucible. The crucible was then kept on a hot plate at a temperature of 120°C until the lubricating oil gets completely dried up. Thereafter, the crucible was kept in a muffle furnace for 4 h at a temperature of 450°C and then for 2 h at 650°C. The ash that remained in crucibles was dissolved in 1.5 ml of HCl solution. The solution was then diluted with 100 ml of deionized water and stored in plastic bottles in a refrigerator at a temperature of 10°C to 15°C. This method was followed to prepare all samples for metal analysis by flame AAS. Variation of these elements was found for different engine operating hours.

4.4 RESULTS AND DISCUSSION

4.4.1 FUEL PROPERTIES

Various fuel properties of MO, SRO, MSO, biodiesel obtained from this mixture of oils, and its blends with HSD (B10 and B20) were determined as per the ASTM standards and are summarized in Table 2. It can be seen from this table that the fuel properties of biodiesel are comparable to those of HSD and are well within the latest American (D 6751–02) and European (EN 14214) standards for biodiesel. The MO, SRO, and MSO, however, were found to have much higher values of fuel properties way above any of these standard limits, thus restricting its direct use as a fuel for diesel engines.

4.4.2 ENGINE PERFORMANCE

The performance parameters such as BSFC, BTE, and EGT obtained with B10, B20, and HSD are found to be affected by fuel blend and engine loading and are discussed in the following sections.

TABLE 2: Fuel properties of biodiesel obtained from MO, SRO, mixture of MSO and their biodiesels

Fuel type	Acid value	Density	Kinematic viscosity	Calorific value	Flash point	Pour point	Carbon residue	Ash content	Water content
	(mg KOH/g)	(kg/m³)	(cSt)	(MJ/kg)	(°C)	(°C)	(%)	(%)	(ppm)
HSD	-	812	2.85	42.5	52	−20.0	0.15	0.01	90
MO	26	913	40.92	36.5	233	14.0	1.6	1.13	1,200
MB100	0.46	861	5.38	37.05	168	3.0	0.22	0.01	450
SRO	2.86	912	44.95	36.6	245	15.0	1.50	1.17	1,400
SRB100	0.39	862	5.58	37.0	146	4.0	0.21	0.01	445
MSOª	14.38	912	42.94	36.55	238	14.0	1.60	1.14	1,300
B10	0.27	817	3.02	42.0	63	−18.0	0.15	0.01	126
B20	0.27	821	3.19	41.4	74	−15.0	0.16	0.01	159
B100	0.33	857	4.56	37.02	164	3.5	0.21	0.013	450
ASTM D6751	<0.80	-	1.9 to 6.0	-	>130	-	-	<0.02	<500
EN14214	<0.50	860 to 900	3.5 to 5.0	-	>120	-	<0.30	<0.02	<500
BIS15607	<0.50	860 to 900	2.5 to 6.0	-	>120	-	-	-	<500

MB100, mahua biodiesel; SRB100, simarouba biodiesel; B100, biodiesel obtained from MSO; B10, 10% MSO with HSD by volume basis;
B20, 20% MSO with HSD by volume basis;
ªMSO, mixture of MO and SRO at 50:50 v/v.

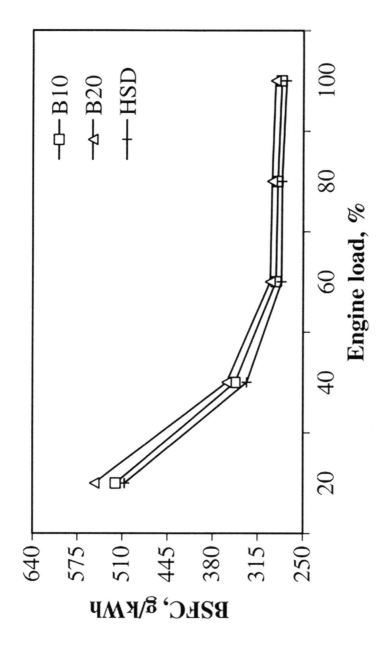

FIGURE 1: Variations of BSFC with engine load for B10, B20, and HSD.

4.4.2.1 BRAKE SPECIFIC FUEL CONSUMPTION

The variations of BSFC with engine load for B10, B20, and HSD are shown in Figure 1. BSFC, in general, was found to increase with an increase in proportion of biodiesel in the fuel blends with HSD. Mean BSFC values with B10 and B20 were found to be 287.29 and 298.67 g/kW h, respectively, and were 2.44% and 5.63% higher than those with HSD. Among the fuel blends tested, 10% blending gave the minimum BSFC, and it increased with the increase in biodiesel percentage in the blends.

BSFC was also observed to decrease sharply with the increase in engine loading for all the fuels tested due to relatively less amount of heat losses at higher loads. At full load, the mean BSFCs for HSD, B10, and B20 were found to be 274.29, 281.05, and 290.00 g/kW h, respectively, as compared with 506.76, 519.52, and 550.37 g/kW h at 20% engine loading. This decrease in BSFC with the increase in engine load might be due to the fact that percentage increase in fuel required to operate the engine was less than the percentage increase in brake power as relatively less portion of the heat losses occurred at higher engine loads.

4.4.2.2 BRAKE THERMAL EFFICIENCY

BTE of diesel engine when operated with HSD, B10, and B20 at different engine loads has been plotted in Figure 2. At 20% engine loading, BTE values were found to be 16.72%, 16.50%, and 15.80% with HSD, B10, and B20, respectively, which were increased to 30.89%, 30.50%, and 29.99% at full load condition. The BTE improved with the engine load for the main reason that a relatively less portion of the power was lost with the increase in engine load. The mean BTE with B10 and B20 was found to be 21.87% and 21.46%, respectively, as compared to 22.27% with HSD. There was a reduction in BTE with the increase in biodiesel percentage in the fuel blends due to the decrease in calorific value of fuel blend.

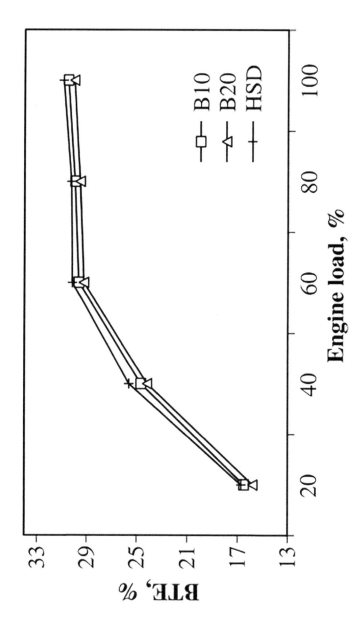

FIGURE 2: Variations of BTE with engine load for B10, B20, and HSD.

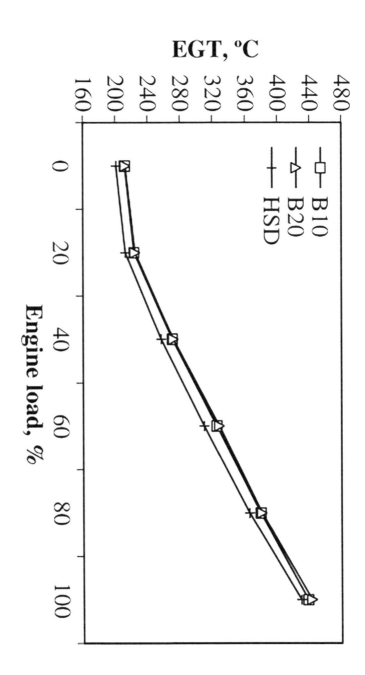

FIGURE 3: Variations of EGT with engine load for B10, B20, and HSD.

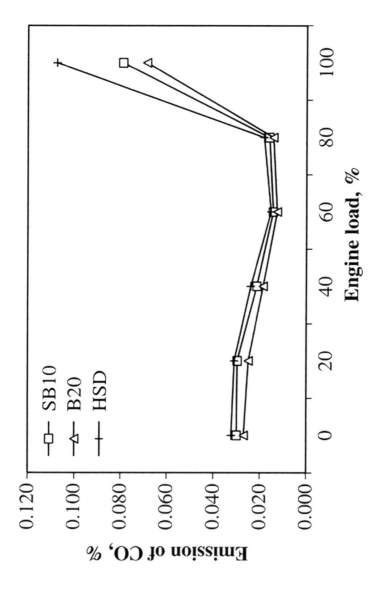

FIGURE 4: Variations of CO emissions with engine load for B10, B20, and HSD.

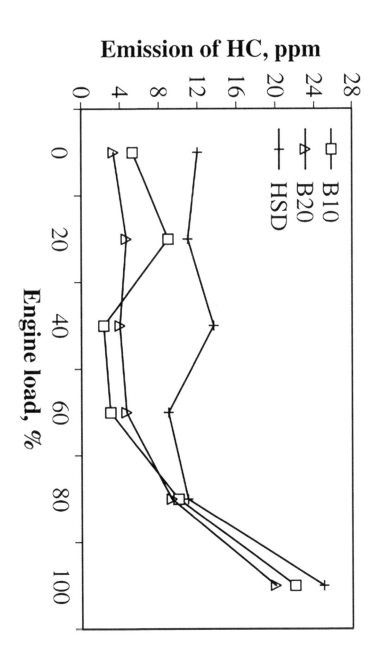

FIGURE 5: Variations of HC emissions with engine load for B10, B20, and HSD.

4.4.2.3 EXHAUST GAS TEMPERATURE

The variations of EGT of a diesel engine when operated with HSD, B10, and B20 at different engine loads are shown in Figure 3. EGT was found to increase with the increase in both concentrations of biodiesel in the blends and engine load. The mean EGT values with B10 and B20 were found to be 308°C and 310°C, respectively, which were 4.44% and 5.12% higher than that with HSD (296°C). The maximum EGT was obtained at full load conditions with all the fuels tested and was 430°C, 438°C, and 443°C with HSD, B10, and B20, respectively, whereas it was 213°C, 224°C, and 225°C at 20% engine loading. The increase in EGT with engine load is obvious from the simple fact that a higher amount of fuel was required in the engine to generate that extra power needed to take up the additional loading.

The variations of BSFC, BTE, and EGT for blends of biodiesel (B10 and B20) obtained from the mixture of oils and HSD at different engine loadings were found to be similar to the trend as reported, while testing biodiesel blends obtained from individual oils (sunflower, linseed, Karanja, rubber seed, rapeseed, soybean, polanga, waste palm, castor, soybean, mahua, and tamanu oils) and HSD in different diesel engines [12–24].

4.4.3 ENGINE EMISSIONS

The average exhaust emissions such as CO, HC, and NOx from diesel engine with different fuels tested were found to be affected by biodiesel blend and engine load and are discussed in the following sections.

4.4.3.1 CARBON MONOXIDE

The variations of CO with engine load when operated with B10, B20, and HSD are presented in Figure 4. CO emission was found to decrease with the increase in proportion of biodiesel in the fuel blends with HSD. The mean values of CO emissions with B10 and B20 were found to be 10.97%

to 21.16% lower than those with HSD. Among the two fuel blends tested, B10 blending gave the maximum CO emission, and it decreased with the increase in biodiesel percentage in the blends.

The mean values of CO emissions for B10 decreased from 0.031% at no-load conditions down to 0.014% at 60% engine load and then increased up to 0.071% at 100% engine load. A similar trend was also observed for B20 and HSD. Initially, at no-load condition, cylinder temperature might be too low, and then, it increased with loading due to more fuel injected inside the cylinder. At an elevated temperature, performance of the engine improved with relatively better burning of the fuel, resulting in decreased CO. However, on further loading, the excess fuel required led to the formation of more smoke, which might have prevented the oxidation of CO into CO_2, consequently increasing the CO emissions sharply.

4.4.3.2 HYDROCARBONS

The HC emissions from the diesel engine when operated with B10, B20, and HSD at different engine loadings are plotted in Figure 5. The mean values of HC emissions with all the fuels tested initially decreased from no load to 60% engine load, and then it increased with further increase in engine load up to 100%. It followed the same trend as that of CO emissions. The mean HC emissions with B10 and B20 were found to be 38.76% and 47.60% lesser than those with HSD. Among the two fuel blends tested, B10 gave the minimum reduction in HC emission as compared with HSD.

4.4.3.3 OXIDES OF NITROGEN

The NOx concentration in exhaust gas from a diesel engine when operated with HSD, B10, and B20 at different engine loadings is shown in Figure 6. It increased with the increase in both engine load and biodiesel concentration in the blends. The maximum NOx values were obtained at full load conditions and were 1,173, 1,211, and 1,230 ppm, respectively, with HSD, B10, and B20, whereas it was 265, 296, and 344 ppm at 20% engine loading. The mean NOx values with B10 and B20 were found to be

667.5 and 693.17 ppm, respectively, and were 5.57% and 11.45% higher than those with HSD. This higher NOx production with biodiesel blends could be attributed to the higher EGT and the fact that biodiesel had some oxygen content in it, which facilitated NOx formation. As the engine load increased, the overall fuel-to-air ratio supplied to the engine increased, resulting in an increase in average gas temperature in the combustion chamber; hence, NOx formation, which is sensitive to temperature, increased.

Similar findings on exhaust gas emissions were observed while operating the diesel engines with different biodiesel obtained from individual oils [23].

4.4.4 SELECTION OF SUITABLE BIODIESEL BLEND

Based on the minimum BSFC, maximum BTE, and minimum EGT, the biodiesel blend, B10, was found suitable for the running diesel engine without compromising engine performance when operated with HSD. However, from the emission point of view, this blend (B10) was found to produce higher CO and HC and lower NOx emissions as compared with B20. However, both fuel blends tested exhibited lower CO and HC and higher NOx emissions as compared with HSD. Considering both performance and emissions, B10 was considered as the suitable blend for replacing HSD in diesel engine and also in the long-term test of the engine that was conducted.

4.4.5 SOOT DEPOSITS ON IN-CYLINDER ENGINE COMPONENTS

The amount of soot deposits on the in-cylinder engine components when operated with different fuels were measured as per the procedure outlined in the 'Estimation of soot deposits on in-cylinder engine components' section and are summarized in Table 3.

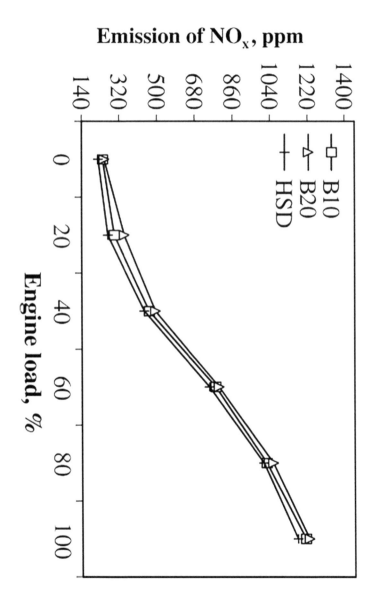

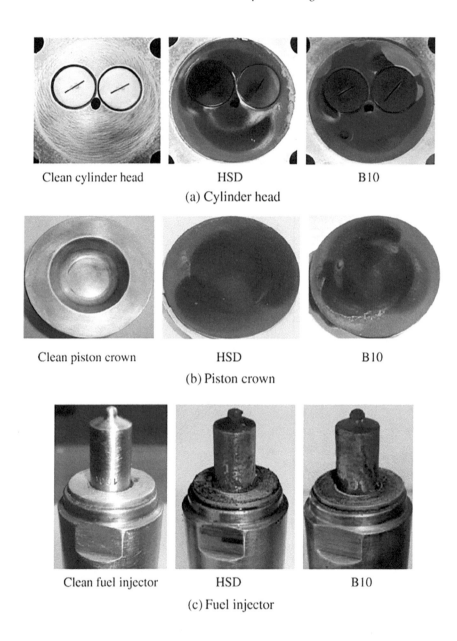

Clean cylinder head   HSD   B10

(a) Cylinder head

Clean piston crown   HSD   B10

(b) Piston crown

Clean fuel injector   HSD   B10

(c) Fuel injector

FIGURE 7: Soot deposits on (a) cylinder head, (b) piston crown, and (c) fuel injector of diesel engine.

TABLE 3: Soot deposits on cylinder head, piston crown and fuel injector of an engine

Engine parts	Operated with HSD (g)	Operated with B10 (g)	Variation as compared to HSD (%)
Soot deposits on cylinder head	0.38	0.23	−39.47
Soot deposits on piston crown	0.52	0.48	−7.69
Soot deposits on fuel injector	0.04	0.03	−25.00

The photographs of soot deposits formed on the cylinder head, piston crown, and fuel injector of diesel engine when operated with HSD and B10 along with a clean cylinder head, piston crown, and fuel injector are shown in Figure 7a,b,c, respectively. From these figures, it can be clearly seen that the soot deposits on the cylinder head, piston crown, and fuel injector of the engine were substantially lower when operated with B10 as compared with HSD. The amount of soot deposits on the cylinder head, piston crown, and fuel injector in the case of the B10-fueled engine was found to be 39.5%, 7.7%, and 25.0% lesser, respectively, as compared with those of the HSD-fueled engine.

A lesser amount of soot deposits on the different engine components when operated with B10 was because of the complete combustion of fuel due to the availability of extra oxygen in biodiesel molecules. The above results are in line with the findings reported by Agarwal et al. [25] when they conducted the 512-h endurance test with 20% blend of linseed oil methyl ester with HSD. They found about a 40% reduction in carbon deposits on the engine components when operated with B20 as compared with HSD.

4.4.6 ADDITION OF WEAR METALS IN LUBRICATING OIL

Addition of wear metals in lubricating oil after 100 h of engine operation each with HSD and B10 was determined by following the procedure outlined in the 'Estimation of soot deposits on in-cylinder engine components' section and is summarized in Table 4. From this table, it can be seen that the concentrations of heavy metals such as Cu, Zn, Fe, Pb, Mg, and Al, except for Mn and Ni, were lower in the lubricating oil of engine when fueled with B10 as compared with when it was fueled with HSD. This could be due to the lesser friction of engine components because of additional lubricating property and presence of fatty acid compounds such as fatty acid methyl esters, FFA, monoglycerides, etc. in B10 fuel. According to some researchers, the presence of long-chain molecules, degree of unsaturation, and oxygenated moieties in the biodiesel play an important role in improving its lubricity [2, 26, 27]. Another reason for this could be the higher viscosity of B10 blend (5.96% higher) as compared with HSD.

Similar findings on the addition of metals on lubricating oil were also reported while conducting 100-and 512-h tests in diesel engines using palm kernel oil biodiesel blends (B7.5 and B15) and linseed oil methyl ester blend (B20), respectively [2, 25].

4.5 CONCLUSIONS

The following conclusions have been made in this study:

1. The fuel properties of biodiesel obtained from the mixture, MSO, were found to be within the limits specified by the biodiesel standards ASTM D 6751–03, DIN EN 14214, and BIS 15607. The fuel properties of biodiesel blends approached those of HSD with a decrease in concentration of biodiesel in the blends.
2. BSFC and EGT of the 10.3-kW diesel engine when operated with biodiesel blends as compared with HSD at different engine loads were found to increase by 2.49% to 5.62% and 4.44% to 5.2%, respectively, whereas BTE was found to decrease by 1.48% to 3.22% with an increase in biodiesel concentration in the fuel blends. Among the two fuel blends tested, B10 had lower mean BSFC (287.29 g/kW h) as well as EGT (308°C) and a higher mean BTE (21.87%).

TABLE 4: Wear metal addition in lubricating oil of engine when operated with HSD and biodiesel blends

Metals	Concentration at 0 h (ppm)	Concentration after 25 h (ppm)		Variation (%)	Concentration after 50 h (ppm)		Variation (%)	Concentration after 75 h (ppm)		Variation (%)	Concentration after 100 h (ppm)		Variation (%)
		HSD	B10		HSD	B10		HSD	B10		HSD	B10	
Cu	0.8	2.8	2.4	-14.3	4.4	3.0	-31.1	5.5	3.9	-29.1	7.02	4.7	-33.0
Zn	915.1	951.1	945.0	-0.64	1,027.8	1,012.0	-1.5	1,063.4	1,042.2	-2.0	1,260.0	1,120.8	-11.0
Fe	17.8	33.7	28.5	-15.4	43.6	37.7	-13.5	55.3	44.4	-19.7	60.9	48.6	-20.2
Mn	1.1	1.4	1.5	7.1	1.6	1.7	6.3	1.7	1.9	11.8	1.9	2.3	21.1
Ni	0.3	0.4	0.4	0	0.4	0.5	25	0.5	0.5	0	1.1	0.9	-18.2
Pb	1.3	2.2	1.4	-36.3	2.8	1.6	-42.8	3.4	1.9	-47.8	4.5	2.8	-37.8
Mg	50.6	59.8	54.2	-9.4	77.2	55.3	-28.4	89.2	57.2	-35.9	115.8	57.9	-50.0
Al	36.6	38.7	37.8	-2.5	58.5	49.3	-15.7	59.6	53.8	-9.7	66.3	57.5	-13.3

3. With an increase in engine load, the BSFC decreased, whereas both EGT and BTE increased for all the fuels tested. However, BTE for all biodiesel blends as compared with HSD was reduced on average by 2.09% at full load, and it was further reduced to 3.41% at 20% engine loading due to higher losses.

4. The CO and HC emissions of the diesel engine when operated with biodiesel blends as compared with HSD were reduced by 10.97% to 21.16% and 38.76% to 47.6%, whereas NOx emissions increased by 5.57% to 11.45%.

5. Based on the performance (minimum increase in BSFC and EGT and with lesser reduction in BTE) and emissions (minimum reduction in CO and HC and minimum increase in NOx), biodiesel blend B10 was selected for long-term use in diesel engine.

6. As compared with HSD-fueled engine, lesser carbon deposits on the in-cylinder parts (such as cylinder head, piston crown, and fuel injector) were observed for the B10-fueled engine due to better combustion of biodiesel blend.

7. Lower concentrations of all heavy metals (such as Cu, Zn, Fe, Pb, Mg, and Al, except for Mn and Ni) in the lubricating oil of diesel engine were found inB10-fueled engine as compared with those in HSD-fueled engine. This could be due to the lesser friction of engine components because of additional lubricity and higher viscosity of the B10 blend as compared with HSD.

Findings of this study would encourage the use of biodiesel obtained from the mixture of oils and would help in reducing the dependency on a particular oil for biodiesel production.

REFERENCES

1. Schumacher LG, Borgelt SC, Hires WG: Fueling a diesel engine with methyl ester soybean oil. Trans. ASAE 1995,11(1):37–40.

2. Kalam MA, Masjuki HH: Biodiesel from palm oil- an analysis of its properties and potential. Biomass Bioenergy 2002,23(6):471–479.

3. Agarwal AK, Bijwe JL, Das M: Wear assessment in a biodiesel fueled compression ignition engine. J. Eng. Gas. Turbines Power 2003, 25:820–826.

4. Schumacher LG, Soylu S, Gerpen JV, Wetherell W: Fueling direct injection diesel engines with 2% biodiesel blends. Appl. Eng. Agric. Trans. ASAE. 2005,21(2):149–152.

5. Schumacher LG, Peterson CL, Gerpen JV: Engine oil analysis of biodiesel fueled engines. Appl. Eng. Agric. Trans. ASAE. 2005,21(2):153–158.

6. Ramaprabhu RO, Bhardwaj M, Abraham : Comparative study of engine oil characteristics in utility vehicles powered by turbocharged direct injection diesel engines fuelled with biodiesel blend and diesel fuel. SAE 2008, 2008280117.

7. Jadhav SS: Development of a biomass based hybrid power generating system. M.Tech. thesis, Indian Institute of Technology, Kharagpur; 2009.

8. Hugo HV, Patricio GVD, Armando G, Héctor P: The development impact of biodiesel: a review of biodiesel production in Mexico. Int. J. Energ. Environ. Eng. 2011, 2:91–99.

9. Jena PC, Raheman H, Prsanna Kumar GV, Machavaram R: Biodiesel production from mixture of mahua and simarouba oils with high free fatty acids. Biomass Bioenergy 2010,34(2):1108–1116.

10. Bureau of Indian Standards: Methods of tests for internal combustion engines: part 8 performance tests. BIS, New Delhi; 1980.

11. Bureau of Indian Standards: Methods of tests for internal combustion engines: part 9 endurance test. BIS, New Delhi; 1980.

12. Moreno F, Munoz M, Morea-Roy J: Sunflower methyl ester as a fuel for automobile diesel engines. Trans. ASAE. 1999,42(5):1181–1185.

13. Agrawal AK, Das LM: Biodiesel development and characterization for use as a fuel in compression ignition engines. Trans. ASME. 2001,123(2):440–447.

14. Raheman H, Phadatare AG: Diesel engine emissions and performance from blends of karanja methyl ester and diesel. Biomass Bioenergy 2004,27(4):393–397.

15. Puhan S, Vedaraman N, Sankaranarayanan G, Bharat Ram BV: Performance emission study of mahua oil (Madhuca indica oil) ethyl ester in a 4-stroke natural aspirated direct ignition diesel engine. Renew. Energ. 2005, 30:1269–1278.

16. Ramadhas AS, Muraleedharan C, Jayaraj S: Performance and emission evaluation of a diesel engine fueled with methyl esters of rubber seed oil. Renew. Energ. 2005,30(12):1789–1800.

17. Labeckas G, Slavinskas S: The effect of rapeseed oil methyl ester on direct injection diesel engine performance and exhaust emissions. Energ. Convers. Manage. 2006,47(13–14):1954–1967.

18. Canakci M: Combustion characteristics of a turbocharged DI compression ignition engine fueled with petroleum diesel fuels and biodiesel. Bioresour. Technol. 2007,98(6):1167–1175.

19. Raheman H, Ghadge SV: Performance of diesel engine with biodiesel at varying compression ratio and ignition timing. Fuel 2008,87(12):2659–2666.

20. Sahoo PK, Das LM, Babu MKG, Naik SN: Biodiesel development from high acid value polanga seed oil and performance evaluation in a CI engine. Fuel 2007, 86:448–454..

21. Ozsezen AN, Canakci M, Turkcan A, Sain C: Performance and combustion characteristics of a DI diesel engine fueled with waste palm oil and canola oil methyl esters. Fuel 2009,88(4):629–636.

22. Nabi MN, Akhter MS, Shahadat MMZ: Improvement of engine emissions with conventional diesel fuel and diesel-biodiesel blends. Bioresour. Technol. 2006, 97:372–378.

23. Gumus M, Kasifoglu S: Performance and emission evaluation of a compression ignition engine using a biodiesel (apricot seed kernel oil methyl ester) and its blends with diesel fuel. Biomass Bioenergy 2010, 34:134–139.

24. Raj MT, Kandasamy MKK: Tamanu oil - an alternative fuel for variable compression ratio engine. Int. J. Energ. Environ. Eng. 2012, 3:1–8.

25. Agarwal AK, Bijwe J, Das LM: Effect of biodiesel utilization of wear of vital parts in compression ignition engine. Trans. ASME 2003, 125:604–611.

26. Geller DP, Goodrum JW: Effects of specific fatty acid methyl esters on diesel fuel lubricity. Fuel 2004, 83:2351–2356.

27. Knothe G, Steidley KR: Lubricity of components of biodiesel and petrodiesel: the origin of biodiesel lubricity. Energy Fuel 2005, 19:1192–1200.

CHAPTER 5

Comparative Analysis of Exhaust Gases from MF285 and U650 Tractors Under Field Conditions

GHOLAMI RASHID, RABBANI HEKMAT, LORESTANI ALI NEJAT, JAVADIKIA PAYAM, AND JALILIANTABAR FARZAD

5.1 INTRODUCTION

Air pollution is a serious problem in all over the world. Diesel engines seem to have a large influence on air pollution because they are used for heavy-duty trucks and emit a higher level of pollutants than petrol engines do, however diesel fuel has slightly higher energy content than petrol per unit volume. Off-road vehicles, trucks, buses and other types of heavy-duty vehicle are powered almost exclusively by diesel engines. Diesel engines make a significant contribution to air pollution (Lindgren and Hansson, 2002).

Comparative Analysis of Exhaust Gases from MF285 and U650 Tractors Under Field Conditions. © *Rashid G, Hekmat R, Nejat LA, Payam J, and Farza J.* Agricultural Engineering International: CIGR Journal *15,3 (2013). http://www.cigrjournal.org/index.php/Ejounral/article/viewFile/2363/1770. An open access article.*

Many studies have been conducted concerning diesel emission analysis and reduction techniques (Jacobs and Assanis 2007; Felsch et al., 2009; Teraji et al., 2009). Pollutants from diesel engines can be roughly divided into groups (Heywood, 1988). The first one is NOx. NOx mainly consists of nitrogen oxide (NO) and nitrogen dioxide (NO_2). The concentration of NO in diesel exhaust is higher than that of NO_2; however, NO_2 has much higher toxicity than NO does. In addition to these two species, N_2O has been recently gathering attention because of its 200 times higher impact factor than carbon dioxide on global warming (http://www.Ipcc nggip. iges.or.jp/public/). Although it can be said that NO, NO_2, and N_2O have different impacts on the environment. The most studies of diesel engine exhaust introduce them as the same species, which is named just NOx.

The second element of diesel exhaust is hydrocarbons and CO. Hydrocarbons consist of thousands of species, such as alkanes, alkenes, and aromatic. Although their toxicity, carcinogenicity, and impact of oxidant formation vary from species to species, they are usually treated together as total hydrocarbons (THC) (ACGIH Website). These uniform treatments of NOx and THC have arisen for two reasons. The first one is that the exhaust gas of automobiles is regulated only by levels of NOx and THC. Another one is the difficulty in measurement. Usually, an analysis of engine exhaust is performed by gas chromatography–mass spectrometry (GC–MS) (Borras et al., 2009). However, achieving quantitative analysis takes a long time. Real time measurement is desirable for engine exhaust analysis because the exhaust gas composition changes in real time along with changes in the engine operating conditions. However, Performing GC-MS in real time is difficult. For these reasons, only a few studies were done about the details of exhaust gas compositions and the effects of engine operating conditions on the compositions (Schulz et al., 1999; Gullett et al., 2006).

The last element of diesel exhaust is particulate matter (PM), which is important to diesel engine exhaust. PM is usually measured by weighing a filter which was exposed to exhaust gas and trapping PM. In a study it is suggested that Nano-particles, generally having a diameter of less than 100 nm although there are different definitions, are more hazardous to human health than larger particles. The standard filter weighing method is regarded as less sensitive for such small particles. According to this reason, the European Commission decided to adopt a new PM measurement technique for automobiles. It is the number counting method, in which the numbers of particles from 23 nm to 2.5 μm are counted. This method has a higher sensitivity to small particles than the standard filter weighing method, because small particles and large particles are treated equivalently (Kittelson, 1998). This discussion indicates that it is important to know the size distribution of particle emissions.

According to an analysis by the Health Effects Institute(HEI, 1995), the composition of diesel exhaust varies considerably depending on engine type and operation conditions, fuel, lubricating oil, and whether an emissions control system is present.

An important proportion of the diesel engine emissions causing environmental problems is caused by work machinery such as agricultural tractors and forestry machines (Hansson et al., 2001). Exhaust emissions from agricultural tractors have a detrimental impact on human health and the environment. In order to reduce these emissions, standards have been introduced and are continuously being tightened (EU 2000, 2004, 2005; Larson and Hansson, 2011). These standards only concern existing vehicles. The latest studies (Hansson et al., 1999) have shown that emission values for agricultural tractor operations cannot be reasonably accurately calculated from average emission factors without account being taken of the type of load on the engine in the operation performed. Large variations in emissions were found between different operations, even when the emissions were related to the mechanical energy output of the engine (emissions in g kWh^{-1}).

Further, there is a great need for precise data on total emission production, for example in processes deciding emission rights and pollution taxes. It is therefore very important that average loads on engines can be decided with high precision. With good knowledge of the average load on a tractor engine, it will also be possible to optimize the engine more effectively in order to minimize the total emissions. The previously described large differences in emission values among different operations indicate that the typical use of the tractor may have important effects on average emission factors. The average use of tractors is related to many more factors than engine size, for example, farm size and its production policy, and the size and age of the other tractors on the farm (Biller and Olfe, 1986).

The largest tractor on the farm is normally used for the heavier operations, e.g. soil tillage, while the other tractors are more often used in lower load operations. It is, therefore, also important to investigate whether the average emission factors differ between tractors of the same size, but with different average use (Hansson et al., 2001).

Recently, many researchers focused on exhaust gases of diesel and petrol engine, such as a comparison between different methods of calculating average engine emissions for agricultural tractors evaluated by Hansson et al. (2001). Environmental impact of catalytic converters and particle filters for agricultural tractors determined by life cycle assessment investigated by Larsson and Hansson (2011), simulated farm fieldwork, energy consumption and related greenhouse gas emissions in Canada by Dyer and Desjardins (2003), effects of vehicle type and fuel quality on real world toxic emissions from diesel vehicles were investigated by Nelson et al. (2008). Emissions from heavy-duty vehicles under actual on-road driving conditions were measured by Durbin et al. (2008). Detailed analysis of diesel vehicle exhaust emissions, nitrogen oxides, hydrocarbons and particulate size distributions were obtained by Yamada et al. (2011). Gaseous and particulate emissions from rural vehicles in China were measured by Yao et al. (2011), etc.

Therefore, the purpose of this work was to measure average some exhaust gases values such as hydrocarbon (HC), carbon monoxide (CO), carbon dioxide (CO_2), oxygen (O_2) and nitrogen oxide (NO) from different tractors at three operations conditions. Engine oil temperature was measured too.

5.2 MATERIALS AND METHODS

The aim of this study is the measurement and correlation between some exhaust gases from two common and popular tractors (MF285 and U650) in IRAN at three situations (use of ditcher, plowing and cultivator operations). The data was recorded in the field with 6,500 m^2 area and clay-loamy soil. Tests by every implement were done in four replications. The operations were done at autumn tillage.

FIGURE 1: Implements

a. Moldboard plow

b. Ditcher

c. Cultivator

Table 1 shows specifications of the tractors was used. The operation conditions were considered for this study and instruments specifications were shown in Table 2. As is shown in Figure 1 ditcher, moldboard plow and cultivator were used. The tractor speed in every operation selected from standard method (ASAE D497.4 MAR99).

TABLE 1: Specifications of the Tractors

	Class	Model	Number of cylinders	Fuel type	Engine operating process	Engine power /hp
Tractor 1	MF 285	1984	4	Diesel	4 - Stroke	75
Tractor 2	U650	1985	4	Diesel	4 - Stroke	65

TABLE 2: Specifications of the instruments and operation conditions

Types of instrument	Characteristics	Engine speed/r min^{-1}	Tractors speed/km h^{-1}
Ditcher	Moldboard ditcher	2250	8
Plowing	Three bottom moldboard plow	220	5
Field Cultivator	Nine tine cultivator	2250	8

FGA-4100 automotive emission analyzer made in China was used for measurement of exhaust gases and engine oil temperature. The details of specific device are illustrated in Figure 2. As is shown in Figure 2 exhaust gases enter to five gas analyzer without dilution. The flow ratio of the exhaust gases is changed by changing the engine speed, so the dilution ratio varied with changing engine speed. Patterns of driving could affect vehicle emissions significantly, so they are very important in measuring vehicle emissions (Hansen et al., 1995; Kean et al., 2003; Yao et al., 2007). Therefore, for solving this problem, engine speeds stabilized during operations with hand accelerator. Then engine speed stabilized at 2,250 r min^{-1}.

1. Gas analyzer

2. Gas inlet

3. Oil temperature
   sensor

FIGURE 2: Details of specific device

Fuel and lubricating oil were constant in both of tractors at every operation, because these parameters are effective on engine emission (HEI, 1995). Both tractors are equipped with the same instruments and worked at the same conditions and measured exhaust gases.

5.3 RESULTS AND DISCUSSIONS

In this research we focused on details of exhaust gases from two common tractors that are used in IRAN at three operation conditions. All of the obtained emission results are presented in Table 3, Table 4 and Table 5. As is shown in these tables measured CO, CO_2 and NO emission from MF285 are higher than U650; however HC and O_2 from U650 are higher than MF285.

TABLE 3: Exhaust gases at ditcher operating.

	MF 285			U650		
	Max	Min	Average	Max	Min	Average
HC/ppm	45	41	43.2	97	65	81.2
CO/%	0.14	0.12	0.13	0.2	0.06	0.11
CO_2/%	8.6	7.5	8.02	3.7	1.1	2.38
NO/ppm	470	271	362.6	105	13	34
O_2/%	12.9	10.8	11.7	18.4	13.8	16.46
Engine oil temperature/°C	68	63.4	65.58	96.3	70.02	70.9

TABLE 4: Exhaust gases at plow operating

	MF 285			U650		
	Max	Min	Average	Max	Min	Average
HC/ppm	49	46	47.75	95	67	82
CO/%	0.27	0.11	0.16	0.2	0.11	0.15
CO_2/%	10.3	7.8	8.92	6.7	3	5.15
NO/ppm	523	369	431	155	43	119.5
O_2/%	11.7	9.92	10.70	17.8	12	13.87
Engine oil temperature/°C	66.5	62.3	64.35	62.3	59.2	60.57

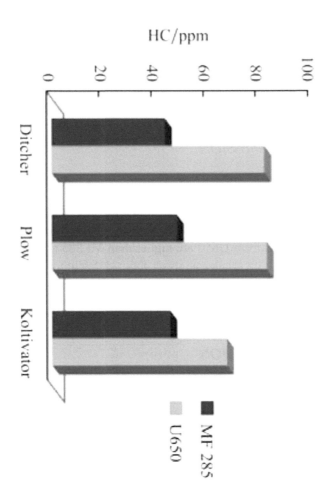

FIGURE 3: HC emission from tractors at operation conditions

TABLE 5: Exhaust gases at cultivator operating

	MF 285			U650		
	Max	Min	Average	Max	Min	Average
HC/ppm	66	41	45.1	76	58	66.75
CO/%	0.16	0.11	0.13	0.1	0.06	0.08
CO2/%	9.8	7.1	8.34	3.8	1	2.2
NO/ppm	560	228	365.8	129	20	68.25
O2/%	13.6	9.08	11.40	18.7	14.4	16.58
Engine oil temperature/°C	54	46.2	51.2	75.3	71	73.77

5.3.1 HC EMISSION

The measured HC values in both tractors in three operating situations are shown in Figure 3. Results showed the value of HC emission while plowing by tractors is higher than the other condition. In addition, the values of exhaust HC from U650 are higher than MF285 at every three operations. The results of variance analysis showed that amounts of exhaust HC have a significant relationship with types of tractors and instrument at 1% as shown in Table 6.

5.3.2 CO EMISSION

The amount of measured gases in both of tractors showed the value of exhaust CO as well as other diesel engines is very low in comparison with petrol engine (DEFRA, ACE Information Programme, Air pollution and Rain Fact sheets Series: KS4 and A). The recorded data showed values of CO in plowing operation are higher than other operations. CO emitted from U650 is lower than MF285 (Figure 4). The results of variance analysis (Table 6) showed that amounts of exhaust CO don't have a significant relationship with types of tractors, but they have significant relationship with types of instrument at 1%. Therefore, values of CO emission are independent from types of tractors.

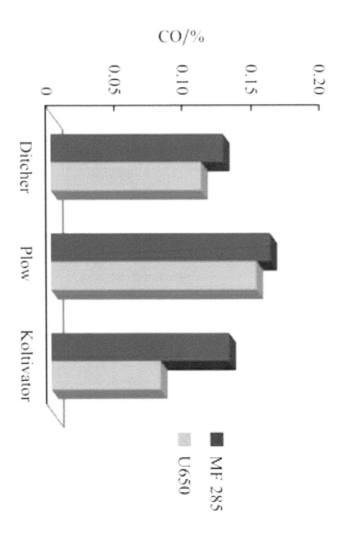

FIGURE 4: CO emission from tractors at operation conditions

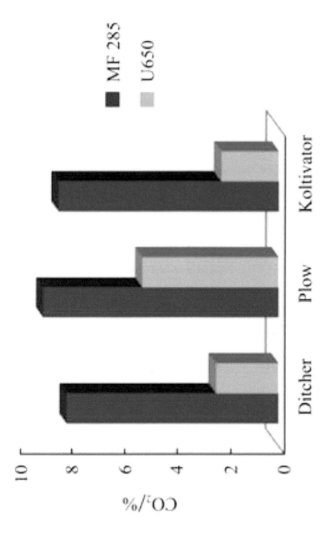

FiGURE 5: CO_2 emission from tractors at operation conditions

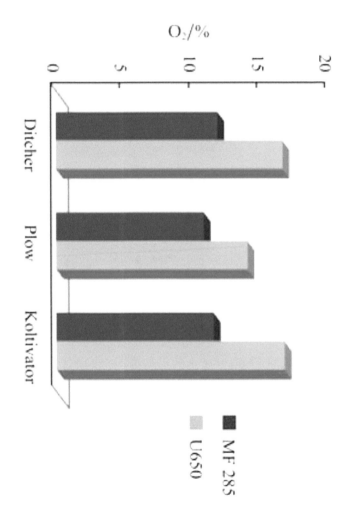

FIGURE 6: O_2 emission from tractors at operation conditions

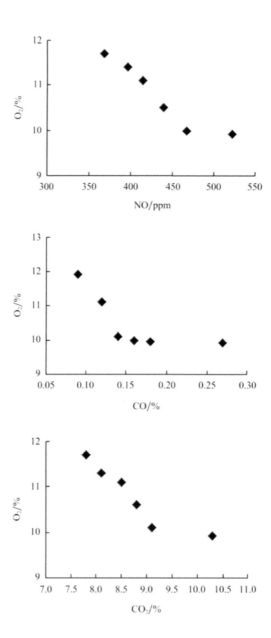

Figure 7: Relationship between exhaust O_2 and other exhaust gases

TABLE 6: Variance analysis

	Tractor			Instrument		
	Mean Square	F	Significant Level	Mean Square	F	Significant Level
HC	1170.40	427.21	0.01>	7.95	0.017	0.01>
CO	0.125	26.89	0.01<	0.006	3.04	0.01>
CO_2	204.34	138.76	0.01>	2.16	0.155	0.01>
O_2	28.64	22.06	0.01>	2.03	0.20	0.01>
NO	139919	105.99	0.01>	2614	0.056	0.01>

The recorded data of CO_2 showed values of this gas at plowing operation are higher than other operations. CO_2 emitted from U650 is lower than MF285 as is shown in Figure 5. The results of variance analysis showed that amounts of exhaust CO_2 have a significant relationship with types of tractors and instrument at 1% (Table 6). Therefore values of CO_2 emission are dependent on types of tractors and instrument.

5.3.4 O_2 EMISSION

The recorded values of O_2 showed amounts of this gas at cultivator operation are higher than other operations and O_2 emitted from U650 is higher than MF285 (Figure 6). Results showed that exhaust O_2 from both of tractors had an inverse relationship with the NO, CO, CO_2 as shown in Figure 7. The results of variance analysis showed that amounts of exhaust O_2 have a significant relationship with types of tractors and instrument at 1% (Table 6). Therefore, values of O_2 emission are dependent on types of tractors and instrument.

5.3.5 NO EMISSION

The experimental data of NO showed values of this gas at plowing operation are higher than other operations (Figure 8). As is depicted in Figure

9, in each tractors NO emission increased as the engine oil temperature increased. Also, the mentioned result was reported for the relationship between NO emission and in-cylinder temperature by Yamada et al. (2011). In addition, NO emission increased by loading enhancement on tractors, therefore NO emission in plow operation was higher than the other conditions. The results of variance analysis showed that amounts of exhaust NO have a significant relationship with types of tractors and instrument at1%. Therefore, values of NO emission are dependent on types of tractors and instrument.

5.3.6 ENGINE OIL TEMPERATURE

The recorded values of engine oil temperature showed amounts of this gas at cultivator operation are higher than other operations (Figure 10).

5.4 CONCLUSION

Evaluation of exhaust gases from diesel engine is important. In this study emission from two current tractors in IRAN at three operation conditions (using ditcher, plowing and cultivator) was reported. Some exhaust gases (HC, CO, CO_2, O_2 and NO) and engine oil temperature were measured.

HC and O2 emission from MF285 were lower than U650, while other emission (CO, CO2, NO) from MF 285 were more than U650. Emission from tractors like other diesel engines, value of exhaust CO is very low in comparison with petrol engines. NO emission increased as engine oil temperature increased. All of exhaust gases except CO have a significant relationship with types of tractor and instruments at 1% as is shown in Table 6. This subject was presented by Durbin et al. (2008), for NO emission from heavy-duty vehicles.

The values of all measured gases at plow operation are higher than the other operations except O_2. It can be clearly seen that amount of exhaust gases depends on amount of loading on tractor. As is shown in Figure 3, Figure 4 and Figure 7 amount of exhaust CO, CO_2 and NO from U650 is lower than MF 285. Therefore used of U650 suggested for agricultural operations, because emission from U650 are lower than MF 285.

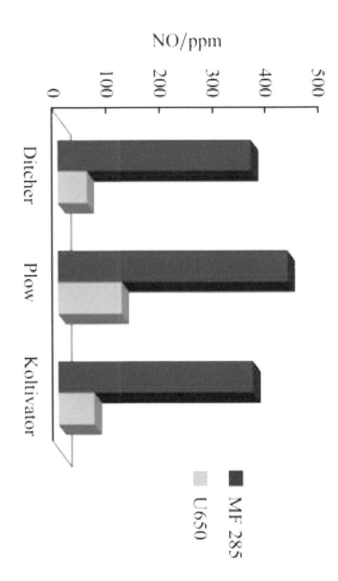

FIGURE 8: NO emission from tractors at operation conditions

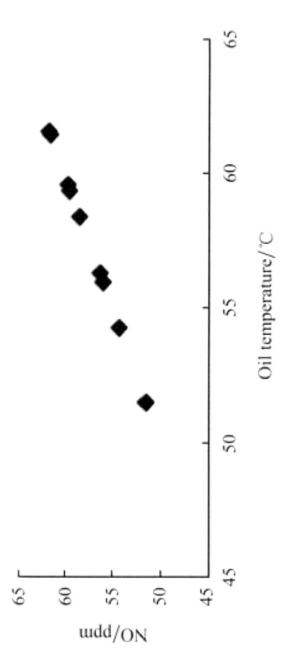

FIGURE 9: Relationship between NO emission and engine oil temperature

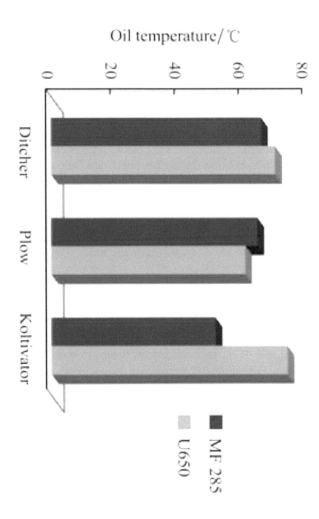

FIGURE 10: Engine oil temperature at operation conditions

REFERENCES

1.   ACGIH Website. 2004. http://www.acgih.org/home.htm. (Accessed August, 2012).
2.   Borras, E., L. A. Tortajad-Genaro, M. Vazquez, and B. Zielinska. 2009. Polycyclic aromatic hydrocarbon exhaust emissions from different reformulated diesel fuels and engine operating conditions. Atmosphere Environment, 43 (37): 5944–5952.
3.   Dyer, J. A., and R. L. Desjardins. 2003. Simulated Farm Fieldwork, Energy Consumption and Related Greenhouse Gas Emissions in Canada. Biosystems Engineering, 85 (4): 503– 513.
4.   DEFRA. Department for Environment, Food and Rural Affairs, ACE Information Programme, Air pollution and Rain Fact sheets Series: KS4 and A).
5.   Durbin, T. D., K. J. Johnson, W. Miller, H. Maldonado, and D. Chernich. 2008. Emissions from heavy-duty vehicles under actual on-road driving conditions. Atmosphere Environment, 42(20): 4812–4821.
6.   EU. 2000. EU, directive /25/EC. European Union Official Journal, L173, 1-34.
7.   EU. 2004. Directive /26/EC. European Union Official Journal, L146, 1-107.
8.   EU. 2005. Commission directive /13/EC. European Union Official Journal, L55, 35-54.
9.   Felsch, C., M. Gauding, C. Hasse, S. Vogel, and N. Peters. 2009. An extended flamelet model for multiple injections in DI Diesel engines. Proceedings of the Combustion Institute, 32(2): 2775–2783.
10. Gullett, B. K., A. Touati, L. Oudejanes, and S. P. Ryan. 2006. Real-time emission characterization of organic air toxic pollutants during steady state and transient operation of a medium duty diesel engine. Atmosphere Environment, 40 (22): 4037–4047.
11. Hansen, J. Q., M. Winter, and S. C. Sorenson. 1995. The influence of driving patterns on petrol passenger car emissions. Science of the Total Environment, 169(1-3): 129-139.
12. Hansson, P. A., O. Nore, and M. Bohm. 1999. Effects of Specific Operational Weighting Factors on Standardized Measurements of Tractor Engine Emissions. Journal of Agricultural Engineering Research, 74(4): 347-353.
13. Hansson, P. A., M. Lindgren, and O. Nore. 2001. A Comparison between Different Methods of calculating Average Engine Emissions for Agricultural Tractors. Journal of Agricultural Engineering Research., 80 (1): 37, 43.
14. HEI. 1995. Diesel Exhaust: A Critical Analyis of Emissions, Exposure and Health Effects. Health Effects Institute, Cambridge, MA, 294pp.
15. Heywood, J. B. 1988. Internal Combustion Engine Fundamentals. McGraw-Hill Education, New York, ISBN: 978-0070286375.
16. IPCC. 2006. http://www.ipcc-nggip.iges.or.jp/public/. (Accessed July 2012)
17. Jacobs, T. J., and D. N. Assanis. 2007. The attainment of premixed compression ignition low-temperature combustion in a compression ignition direct injection engine. Proceedings of the Combustion Institute, 31(2): 2913–2920.
18. Kean, A. J., R. A. Harley, and G. R. Kendall. 2003. Effects of vehicle speed and engine load on motor vehicle emissions. Environ. Sci. Technol, 37(17): 3739-3746.

19. Kittelson, D. B. 1998. Engines and Nanoparticles: A Review. J. Aerosol Sci, 29(5-6): 575–588.
20. Larsson, G., and P. A. Hansson. 2011. Environmental impact of catalytic converters and particle filters for agricultural tractors determined by life cycle assessment. Biosystem engineering, 109 (1): 15-21.
21. Lindgren, M., and P. A. Hansson. 2002. Effects of Engine Control Strategies and Transmission Characteristics on the Exhaust Gas Emissions from an Agricultural Tractor. Biosystems Engineering, 83 (1): 55–65.
22. Niles, Rd., and St. Joseph. ASAE STANDARDS. 2000. The Society for engineering in agricultural, food, and biological systems.
23. Nelson, P. F., A. R. Tibbett, and S. J. Day. 2008. Effects of vehicle type and fuel quality on real world toxic emissions from diesel vehicles. Atmosphere Environment, 42(21): 5291–5303
24. Schulz, H., G. B. De Melo, and F. Ousmanov. 1999. Volatile Organic Compounds and Particulates as Components of Diesel Engine Exhaust Gas. Combustion and Flame, 118(1-2): 179–190.
25. Sullivan, J. L., R. E. Baker, B. A. Boyer, R. H. Hammerle, T. E. Kenney, L. Muniz, and T. J. Wallington. 2004. CO2 emission benefit of diesel (versus gasoline) powered vehicles. Environmental Science and Technology, 38(12): 3217–3223.
26. Teraji, A., Y. Imaoka, T. Tsuda, T. Noda, M. Kubo, and S. Kimura. 2009. Development of a time-scale interaction combustion model and its application to gasoline and diesel engines. Proceedings of the Combustion Institute, 32(2): 2751–2758.
27. UN-GRPE Phase 3 Inter-laboratory Correlation Exercise: Updated Framework and Laboratory Guide for HD engine Testing, A Document for the UK Department for Transport, RD04/201901.3b, 25 January 2006 <http://www.unece. org/trans/ main/ wp29/ wp29wgs/ wp29grpe/ pmp15. html>.
28. Yamada, H., K. Misawa, D. Suzuki, K. Tanaka, J. Matsumoto, and M. Fuji. 2011. Detailed analysis of diesel vehicle exhaust emissions: Nitrogen oxides, hydrocarbons and particulate size distributions. Proceedings of the Combustion Institute, 33(2). 2895–2902.
29. Yao, Z., Q. Wang, K. He, H. Huo, Y. Ma, and Q. Zhang. 2007. Characteristics of real- world vehicular emissions in Chinese cities. J. Air Waste Manage. Assoc., 57(11): 1379-1386.
30. Yao, Z., H. Huo, Q. Zhang, D. G. Streets, and K. He. 2011. Gaseous and particulate emissions from rural vehicles in China: Atmosphere Environment, 45 (18): 3055-3061.

CHAPTER 6

Respiratory Quotient (Rq), Exhaust Gas Analyses, CO_2 Emission, and Applications in Automobile Engineering

KALYAN ANNAMALAI

Development and economic growth throughout the world will result in increased demand for energy. Currently almost 90% of the total world energy demand is met utilizing fossil fuels [1]. Petroleum and other liquid fuels include 37% of the total fossil reserves consumed for transportation and other industrial processes [1]. Emission of harmful gases in the form of nitrogen oxides, sulphur oxides and mercury are the major concerns from the combustion of conventional energy sources. In addition to these pollutants, huge amount of carbon dioxide is liberated into the atmosphere. Carbon dioxide is one of the green house gases which cause global warming. Though technology is being developed to sequester the CO_2 from stationary power generating sources, it is difficult to implement such a technology in non-stationary automobile IC engines.

Respiratory Quotient (Rq), Exhaust Gas Analyses, CO$_2$ Emission and Applications in Automobile Engineering © Annamalai K. Advances in Automobile Engineering *2,2 (2013), http://dx.doi. org/10.4172/2167-7670.1000e116. Licensed under Creative Commons Attribution License, http://creativecommons.org/licenses/by/3.0/.*

Some of the methods which has been studied to reduce the amount of CO_2 being released from the IC engines include blending ethanol with gasoline. Ethanol produced from corn, sugarcane bagasse and lignocellulosic biomass is considered to be carbon neutral. It is assumed that the carbon released during combustion of ethanol will be readily absorbed by the plants and will be recycled and hence the carbon released in such a process will not be accounted in the carbon footprint. But such an approach with biomass is being challenged by a number of recent studies. Land use change, use of electrical energy and fossil based energy for collection, and transportation and processing of biomass, energy conversion efficiency and productivity of forest land impacts the decision on carbon neutrality of biomass based fuels [2]. The use of short rotation woody crops with much higher yield [3] and reduced costs can serve as a source for ethanol production or energy applications.

Recent developments in the IC engines have resulted in reducing the emissions and improving the vehicle efficiency by using different sensors such as oxygen and NOx sensors and utilizing exhaust gas recirculation to lower the oxygen concentration and reduce the temperature within the engines.

The O_2 sensor which operates at about 344°C (650°F) reads 0 volts at lean condition to almost 1.0 volt under rich condition [4]. Typically the ideal A:F for a gasoline engine is 15:1 (by mass). Oxygen sensor serves to maintain a near stoichiometric condition (14.7:1 by mass) and vary the air fuel ratio within the engine for complete combustion of fuel and hence reduce the emission of unburnt hydrocarbons (HC) and CO [5]. Ideal O_2% in exhaust is less than 1.5%. A high amount of HC and CO may indicate problems in combustion and appropriate A: F ratio. x It has been found that the higher heating value per unit mass of oxygen burned (HHVO2) remains approximately the same for all fuels; in fact this approximation is used in biology to determine the metabolic rate (or energy release rate) of humans by measuring O_2 used (=O_2 inhalation rate – O_2 exhalation rate) and using known values of $HHVO_2$. When renewable fuels (e.g. ethanol, C_2H_5OH) are blended with fossil fuels (gasoline, CHx) and used for combustion in IC engines, same thermal energy input is assured under fixed air flow rate and fuel flow is adjusted such that same O_2% is maintained in exhaust. Thus the oxygen sensor in an IC engine helps to maintain a proper air fuel ratio, similar heat input rate (or power) and excess air %. For example the heating value of gasoline and ethanol blend is lower than gasoline and hence blend fuel flow rate must be increased until the O_2% is maintained the same when fuel is switched from gasoline to blend [6,7].

Recently CO_2 based motor vehicle tax has been introduced in European Union countries to promote the utilization of renewable fuels in automobiles [8]. Taxes will be levied to the customer based on the emission of CO_2 in g/km. Environmental transport association (ETA) has proposed an empirical rule to determine the CO_2 emission from gasoline and diesel vehicles [9]. If Miles Per Gallon (MPG) is 40, the empirical rule for the SI engine to determine the CO_2 emission is 6760/MPG=6760/40=169 g of CO_2/km. For Diesel engine: 7440/ MPG=7440/40=186 g/km. Environment Protection Agency (EPA) has estimated the amount of carbon released on combustion of gasoline and diesel to be around 8887 and 10180 grams CO_2/gallon respectively for each of the fuel [10].

Rather than using the empirical rule, the potential of a particular fuel used in automobile IC engine to produce carbon dioxide can be estimated from the fuel ultimate analysis. A term used in biological literature to determine the basal metabolic rate, Respiratory Quotient (RQ) has been used in Ref [7] to estimate the CO_2 emission potential of fuels. RQ is defined as the ratio of the moles of CO_2 emitted to the moles of O_2 consumed typically for oxidation reaction of a fuel.

RQ factor can also be used to estimate the amount of CO_2 being released on burning fossil fuels. Higher the RQ value of fuel higher the potential to emit CO_2 per unit heat input to the IC engine. Apart from using the fuel chemical formula and fundamental combustion literature, exhaust gas analysis from automobiles can also be used to determine the RQ. The RQ CO_2 emission in tons per GJ of energy input can be estimated from the knowledge of RQ of a particular fuel. The CO_2 in tons per GJ of energy input is given as [7].

Estimation of RQ factor from the empirical chemical formula or fuel ultimate analysis can be performed [7].

$$RQ = \frac{CO_2 \; moles \; produced}{O_2 \; moles \; consumed} = \frac{1}{\left\{1 + \left(\frac{H}{4C}\right) - \left(\frac{O}{2C}\right) + \left(\frac{S}{2C}\right)\right\}} \qquad (1)$$

where C, H, O and S are the number of carbon, hydrogen, oxygen and sulphur atoms respectively. Instead of fuel composition, exhaust gas analyses can be used to determine RQ in addition to estimation of air fuel (A:F) ratio used, excess air %, and φ, the equivalence ratio or stoichiometric ratio SR (= 1/φ). If φ<1 or SR>1, it implies lean mixture. General methodology is to formulate the following reaction equation, assume complete combustion and use atom conservation assuming the fuel to be CHxOy where x=H/C and y=O/C

$$CH_xO_y + aO_2 + bN_2 \rightarrow cCO_2 + dH_2O + eO_2 + fN_2$$

(2)

There are 8 unknowns for C-H-O fuel: x, y, a, b, c, d, e and f. Thus one needs 8 equations. Four equations are obtained from an atom balance of C, H, O, and N. The four additional equations are generated as follows. The ratio {b/a} in the intake air is known as 3.76; the percent of O_2, CO_2, and H_2O are known from the exhaled gas composition. One can derive the following formula for RQ from exhaust gas analysis [7]:

$$RQ = \frac{CO_2\ moles\ produced}{O_2\ moles\ consumed} = \frac{X_{CO_2,e}\left(\dfrac{X_{N_2,i}}{1 - X_{O_2,e} - X_{CO_2,e}}\right) - X_{CO_2,i}}{X_{O_2,i} - X_{O_2,e}\left(\dfrac{X_{N_2,i}}{1 - X_{O_2,e} - X_{CO_2,e}}\right)}$$

(3)

Where N_2 X , CO_2 X and O_2 X are the mole fractions of nitrogen, carbon dioxide and oxygen which could be either on dry or wet basis and subscripts i and e refer to inlet and exit of combustion chamber respectively. The fuels gasoline, Diesel and Kerosene have chemical formula to be CHx and as such one needs only 7 equations and hence needs only % of 2 species in exhaust (e.g.: O_2 and CO_2% or O_2% and N_2%). Typically N and S in fuels are trace species and hence one may apply the above analysis even for $CH_xO_yN_zS_s$ fuels. From the exhaust gas data for a gasoline engine [6] (HC: 750 ppm; NOx: 1050 ppm; CO: 0.68%; H_2: 0.23%; CO_2: 13.5%; O_2:

0.51%; H_2O: 12.5%; N_2: 72.5%) and dumping small amounts of H_2 and CO with N_2, , the RQ of the fuel was estimated to be 0.71 (if CO and H_2 remain in exhaust, O_2% =0.51) to 0.73 (if CO and H_2 are burnt to CO_2, H_2O, O_2% reduced to 0.055) using Equation (2). A set of formulas were derived depending upon available exhaust species % to determine the RQ from available exhaust gas analysis data. EXCEL based software had been developed to estimate all unknown in Equation (2). CO_2 emission in tons per GJ of energy input can be estimated from the knowledge of RQ of a particular fuel. The CO_2 in tons per GJ of energy input is given as [7]

$$CO_2 \text{ in tons per GJ energy input} \approx RQ * 0.1 \qquad (4a)$$

$$CO_2 \text{ in short tons per mmBTU} \approx RQ * 0.116 \qquad (4b)$$

If one uses the fuel composition, RQ value for motor gasoline ($CH_2.02$ with C mass %=84.5, [14]) was found to be 0.66 (or 0.066 tons of CO_2 per GJ heat input); the RQ for diesel is 0.68, and biodiesel 0.70 and alcohols have a RQ of 0.67 [7]. Note that RQ seems to fluctuate from 0.66 to 0.73 even for gasoline and as such exhaust gas analyses provide more reasonable values of RQ during operation of automobiles. More details are given in [7].

An empirical formulae for flue gas volume is given in ref 13 and it can be used to estimate NOx in kg/GJ. Thus

$$NOx \text{ in g of } NO_2/GJ = \{NOx \text{ in ppm}\} *1.88 \times 10^{-3} *Flue \text{ gas volume in } m^3/GJ \quad (5)$$

$$Flue \text{ gas volume in } m^3/GJ = \{3.55+0.131\ O_2\% + 0.018 * (O_2\%)2\}\ (H/C)2- \{27.664 + 1.019\ O_2\% + 0.140* (O_2\%)2\}\ (H/C) + \{279.12 + 10.285\ O_2\% + 1.416* (O_2\%)2\} \qquad (6)$$

Using exahust gas analyses presented before, O_2%=0.51% (φ=0.97), the estimated flue gas SATP volume is 246 m^3/GJ. With NOx=1050 ppm, the formula (5) yields 480 g/GJ.

For NO emissions, the RQ and X CO_2,e in exhaust can also be used to convert the NO in ppm into g/GJ or lb per mmBTU[11]. See Ref [13] for reporting emissions in different forms

$$NO \text{ in } g \text{ per } GJ = 0.102* NO \text{ in } ppm* \{RQ/X\ CO_2,e\} \tag{7}$$

Using Equation. (7), X CO_2,e = 0.135, RQ=0.71, NO= 1050 ppm, the NO in g/GJ is 496 g/GJ. As XO_2e increases (lean mixture), X CO_2e decreases, NO in g per GJ increases for same NO in ppm due to higher mole or volume flow rate of exhaust gases for the same energy released.

It should also be noted that the production of renewable fuels such as ethanol will consume some fossil resources which will emit CO_2. Hence, in addition to RQ for oxidation, one must define an equivalent RQprocess which should take into account the amount of CO_2 released during processing of the fuel (e.g. ethanol production from corn) and

$$Net\ RQ = RQ + RQprocess$$

Equations (4) can be modified to determine the total amount of CO_2 emitted per unit distance travelled by an automobile while consuming different fuels [7]. With heat value of 123,361 BTU/gallon (34,383 kJ/L or 45844 kJ/kg assuming ρ=750 kg/m³ for gasoline, Equations (4) transform to

$$CO_2 in \left(\frac{lb}{mile}\right) = 28.62 * \left(\frac{RQ}{MPG}\right) \tag{8a}$$

$$CO_2 in \left(\frac{kg}{km}\right) = 3.44 * \left(\frac{RQ}{km\ per\ L}\right) \tag{8b}$$

Using Equations (8) the amount of CO_2 emitted on using gasoline were estimated to be around 0.47 lb per mile or 0.13 kg per km assuming 40 MPG (16.9 km per L). Net CO_2 emitted for a blend of 85:15 (vol. %) gasoline: Ethanol was 0.12 kg /km or 0.42 lb per mile(including CO_2 from both gasoline and ethanol) assuming same 40 MPG [7]. Two points need to be noted for blends: Firstly, the MPG will not be the same for the blend (CH_xO_y) since the amount of energy in a gallon will be less for the blend. Hence Equation (4) which gives the CO_2 emission in kg/GJ would be a better representation for determining the CO_2 emission, taxing the vehicle and ranking fuels based on CO_2 emitting potential. Secondly one must not include CO_2 from ethanol oxidation in gasoline: ethanol blend since ethanol is renewable fuel; excluding CO_2 from ethanol, the CO_2 in kg per km is reduced to 0.11 kg per km or 0.39 lb per mile.

With recent developments in sensor technology and engine controls, the data from the gas sensors at the engine tail pipe can be used to keep track of the total CO_2 being emitted while driving an automobile. The formulas presented here can be included in an algorithm in the engine Onboard Diagnostics (OBD) to calculate the cumulative CO_2 emitted from an automobile during its life time. Including this number in an automobile dashboard will enable continuous monitoring and logging of CO_2. With ongoing research to increase the production of short rotation woody crops for ethanol production, more efficient ways of converting biomass to renewable liquid fuels will be identified and implemented.

REFERENCES

1. http://www.eia.gov/oiaf/aeo/tablebrowser/#release=IEO2013&subject=0-
2. Johnson E Goodbye to carbon neutral: Getting biomass footprints right. Environ. Impact. Assess. Rev. 29(2009), 165-168
3. Keoleian GA, Volk TA (2005) Renewable energy from willow biomass crops: life cycle energy,   environmental and economic performance. BPTS 24: 385-406.
4. Chen W, Annamalai K, Ansley RJ, Mirik M, "Updraft Fixed Bed Gasification of Mesquite and Juniper Wood Samples. J of Energy 41:454-461.
5. Ramamoorthy R, Dutta PK, Akbar SA (2003) Oxygen sensors: materials, methods, designs and applications. Journal of materials science. 38: 4271-4282.
6. Annamalai K, Thanapal SS, Lawrence B, Ranjan D Respiratory Quotient (RQ) in Biology and Scaling of Fossil and Biomass Fuels on Global Warming Potential (GWP). Renewable Energy, under review, 2013

7. http://www.ecocatalysis.com/en/articles/Automotive-exhaust.html, accessed August 9, 2013.

8. Annamalai K, Thanapal SS, Lawrence B, Ranjan D (2013) Respiratory Quotient (RQ) in Biology and Scaling of Fossil and Biomass Fuels on Global Warming Potential (GWP).

9. https://www.eta.co.uk/2010/02/22/calculating-a-cars-co2-emissions-from-its-mpg/, accessed August 9, 2013. www.epa.gov/otaq/ , accessed Aug 8, 2013.

10. PB , " W y A F R S ?" 46, Motor, December 2005; also access www.motor.com

11. Annamalai K, Puri I K (2007) Combustion Science and Engineering CRC Press.

12. North American Combustion Handbook, Vol I 3rd Ed, 2001

PART IV

EMISSIONS FROM FUEL
AND ELECTRICITY PRODUCTION

CHAPTER 7

Life Cycle Greenhouse Gas Emissions from Electricity Generation: A Comparative Analysis of Australian Energy Sources

PAUL E. HARDISTY, TOM S. CLARK, AND ROBERT G. HYNES

7.1 INTRODUCTION

Providing the benefits of electricity to hundreds of millions of people around the World is a key challenge of this century. In the International Energy Agency's World Energy Outlook 2010, global energy demand was expected to rise 1.4% per year on average to 2035, assuming no change in current business-as-usual energy policy [1]. In 2010, actual global energy use jumped by 5.6%, the largest single year increase since 1973 [2]. The current global energy mix remains heavily weighted towards conventional fossil fuels. Coal's share of global energy consumption was 29.6%, the highest since 1970. By 2030, it is expected that World energy consumption

Life Cycle Greenhouse Gas Emissions from Electricity Generation: A Comparative Analysis of Australian Energy Sources, © *Hardisty PE, Clark TS, and Hynes RG.* Energies **5** *(2012), doi:10.3390/en5040872. Licensed under Creative Commons Attribution 3.0 Unported License, http://creativecommons.org/licenses/by/3.0/. Used with permission of the authors.*

will rise from just under 12 btoe (billions of tonnes of oil equivalent) to over 16 btoe, with much of this growth occurring in non-OECD countries, particularly China and India [3].

In line with the rapid growth in energy consumption, and reflecting the current heavy dependence on fossil fuels, global anthropogenic greenhouse gas emissions grew by 5.9% in 2010, the steepest single year increase since 1972. In 2009, worldwide fossil fuel consumption subsidies amounted to $ 312 bn, with oil products and natural gas the largest recipients, at $ 126 bn and $ 85 bn respectively [1].

Such trends are at a time when scientists, economists and government leaders around the world have recognized the need to significantly lower emissions and stabilize atmospheric CO_2 levels to avoid the worst predicted effects of climate change. To this end, the Australian government has introduced legislation which will put a price on carbon emissions by 2012, partially internalizing what heretofore has been an externality for Australians. In doing so, Australia is following in the footsteps of the European Union, Norway, several American states and Canadian provinces, all of whom are applying some mechanism to provide an economic incentive to reduce emissions. As a major exporter of fossil fuels, notably LNG and coal, and one of the highest per capita users of fossil fuels, including brown coal, Australia faces significant challenges both in pricing carbon, and in understanding the effects of such pricing on export markets. Meeting rising power demand while simultaneously driving down global emissions of the greenhouse gases which drive anthropogenic global warming will require clear, accurate information on the relative emissions intensities of power generation options.

A variety of studies are available in the literature, which examine the life-cycle emissions of various fuel types [4–7]. Recent studies in the Australian context have focused on exports to Asia of Northwest Shelf gas (conventional gas), coal seam gas (CSG), and Australian black coal [8–10]. These studies have concluded generally that LNG has lower overall life-cycle GHG emissions than coal, when power generation technologies of similar efficiency or application are compared (e.g., gas from LNG burned in open cycle generation produces 35% less emissions than sub-critical coal-fired technology, for instance). Open cycle gas-fired technology for Australian Northwest Shelf gas LNG produced 41% fewer emissions than the worst (sub-critical) coal technology [8]. Open cycle gas technology, using LNG from CSG, produced 27% and 5% fewer GHG emissions over its life cycle than sub-critical and ultra-supercritical coal fired technology, respectively, burning Australian black coal [9]. CSG was found to be more GHG intensive than conventional Northwest shelf gas, on a like-for-like basis, but this CSG study [9] did not consider upstream fugitive emissions in any detail.

The US Environmental Protection Agency (USEPA) has estimated that worldwide leakage and venting of natural gas (methane) would reach 95 billion m^3 in 2010 [11]. Other recent work from the USA has estimated that fugitive emissions could add as much as 3–6% to the total life cycle emissions for shale gas [12]. This and other work suggests that with application of best practice, fugitive emissions can be significantly reduced. Other work has examined the life cycle GHG emissions of nuclear power and various renewable energy sources [13,14]. None of the existing studies in the Australian context have examined and compared the life cycle GHG emissions of a wider range of power sources such as export fossil fuels, domestic gas, nuclear and renewables.

7.2 APPROACH

This study is based on a review of original source data from public submissions in Australia, available studies in the literature, and the authors' experience. This study focuses on the Australian context, which, as discussed below, differs from the American situation in a number of respects. While in the US gas is used predominantly for heating [12], when Australian gas is exported as LNG, electricity production is the primary use. On this basis, when comparing energy sources, GHG emissions in this paper are estimated and compared based on the functional unit of MWh of electricity sent out from a power station (after efficiency losses). The analysis is an attributional life cycle assessment, based on static, current emissions, and thus is inherently limited in assessing future emissions, especially the impact of innovation and other system changes. For policy making, consequential LCAs involving dynamic modelling can be useful.

In deriving GHG emissions estimates, the Greenhouse Gas Protocol of the World Business Council for Sustainable Development and the World Resource Institute was followed [15]. The Australian Government's National Greenhouse and Energy Reporting methodology is consistent with the Protocol [16]. Estimates were developed following the Australian Government's National Greenhouse and Energy Reporting (NGER) (Measurement) Determination [16]. In the case of fugitives from natural gas operations, latest available studies in the peer-reviewed literature were used to supplement the American Petroleum Institute guidelines (the API Compendium) [17].

All emissions are converted to carbon dioxide equivalents (CO_2-e) as specified under the Kyoto Protocol accounting provisions to produce comparable measures of global warming potential (GWP). The GWP factors used are those specified in the Australian NGA Factors (carbon dioxide 1, methane 21 (over 100 years) and nitrous oxide 310) [18]. The values adopted by the Australian Government are based on IPCC 1995 values [19].

GWPs relative to carbon dioxide change with time as gases decay. The latest estimates for the GWP of methane over 20 years are between 72 [20] and 105 [21]. To provide a conservative view, this study also examines the effect of fugitives using the higher, most recent 20 year GWP of Shindell et al. [21].

7.2.1 GENERAL ASSUMPTIONS

In developing GHG life cycle emissions estimates for a comparative analysis, certain key assumptions are required to normalize the data. For export scenarios, China is assumed to be the destination for comparison, although in practice both Australian LNG and black coal have multiple destinations. There is some piping of gas to individual power stations but, for comparability, power stations are assumed to be at or near the port and pumping energy use is not material.

For the base comparison, emissions from existing technologies are assumed to apply for the comparison, including best practice for GHG mitigation. A normal range of combustion technologies for gas combustion and power generation has been assumed. These technologies are internationally similar for power generation although the mix of types and relative efficiencies (and greenhouse emissions) will vary from country to country. For gas combustion, estimates have been made for open cycle gas turbine (OCGT, average efficiency 39%) and combined cycle gas turbine power plant (CCGT, average efficiency 53%). In practice there is wide variation in efficiencies around these figures. For coal combustion, estimates have been made for sub-critical (average efficiency 31%), supercritical (average efficiency 33%) and ultra-supercritical (average efficiency 41%) pulverized fuel power plant. Again, in practice there is wide variation in efficiencies around these figures.

The timeline for comparison spans from the present, considering technologies currently applied or going on-stream, while considering average emissions over the life of a project. For LNG, CSG and coal projects this is typically up to 30 years. While there may be some technology changes over this time, especially improvements in end-use combustion efficiency, the technologies for both industries are generally well established and most GHG emissions can be readily estimated based on activity levels and other factors.

Estimates include emissions from construction, emissions embedded in materials, production, transport, and from combustion. Fugitive emissions across the life-cycle are also included. When considering the life cycle emissions for renewable and nuclear energy, the vast majority of emissions are related to construction and embedded in materials. Embedded emissions in non-Australian project capital equipment were not included on the grounds of immateriality [22].

7.2.2 ASSUMPTIONS FOR BLACK COAL

Source data from publicly available submissions varied in terms of inclusion of emissions types. While all included diesel use, fugitives and explosives and many use grid power, reporting of other emissions varied. Industry averages were developed from the cases available and included in the base case. Atypical emissions such as gas flaring from underground mines were not included.

There are general differences between open cut and deep (underground) mines, especially in levels of fugitives, relative use of diesel and electricity and, for some underground mines, use of gas for power generation. The analysis reflects these differences, and provides a range of emission intensities. The base case assumes coal from large open cut mines which dominate the export industry. It is assumed that 100% of the gas content of fugitives released is methane.

TABLE 1. Australian black coal: GHG emissions sources.

Operation	Emissions sources
Extraction and processing • Open cut mining operations • Deep mining operations • Preparation plant for all mines includes crushing, screening, sizing, washing, blending and loading onto trucks and conveyors	• Use of diesel for generators (used for plant and equipment) and vehicles • Use of grid electricity for some mines (scope 2 or indirect emissions) • Fugitives (more significant for 'gassy' underground mines) • Use of explosives • Slow oxidation • Spontaneous combustion • Construction emissions • Embedded emissions in materials and fuel
Transport Most coal is transported by rail to port where it is transferred to bulk carriers. Rail shipment distance range from less than 20 km to around 400 km and may be on dedicated or shared systems.	Use of diesel for locomotives (or electricity for electrified railways), electricity in port handling, fuel for ships.
Combustion The most common modern type of power plant in all export and domestic markets is pulverized coal power plant where the coal is pulverized in the receiving power station. Various combustion technologies are commonly employed, including sub-, super and ultra-super critical with various efficiencies in electricity sent out.	The main life cycle emissions arise from the use of coal in power generation, including internal use of power in pulverization and other plant systems (which contributes to efficiency losses).

Spontaneous combustion may occur in stockpiles and release greenhouse gas emissions and estimates are made based on data from environmental impact statements (EIS), and have been included in this analysis. However, there is no accepted international or Australian methodology for estimating this type of emission. Other sources of emissions which have not been included, on the basis of immateriality, include land clearance and offsets from rehabilitation, and waste gas draining and gas flaring from underground mines. For pulverized fuel combustion, the shipped coal is pulverized to the required specification. Power use in crushing mills is part of the internal power use of a power station and is reflected in overall efficiency figures. Pulverization is assumed to take up to 2% of output, and feed pumps and other systems another 2%.

7.2.3 ASSUMPTIONS FOR ALL NATURAL GAS

This analysis considered natural gas exported as LNG from both conventional Northwest Shelf gas and CSG. As noted above, for simplification, it was assumed that the power station at the receiving country is close to port, requiring minimal energy for transmission. Loss of LNG product occurs in shipping (1.5% loss of LNG product cargo as shipping fuel) and in re-gasification (2.7% lost in fueling re-gasification heaters). Where an LNG plant processes condensate and domestic gas, GHG emissions for LNG exports are apportioned. For the LNG base case, production of 10 Mtpa (a 3 train LNG plant) is assumed.

7.2.4 ASSUMPTIONS FOR COAL SEAM GAS

For coal seam gas scenarios, the study considers GHG emissions from the exploration phase, including coreholes and operation of pilot wells, construction and operation of production wells, gas gathering lines, gas compressors and gas dehydration equipment. The base case assumes zero venting in gas field development and operations (i.e., all fugitive emissions are flared). At present, the CSG industry is nascent in Australia, and there is little operational data to support this assumption. However, most CSG proponents have stated in their EIS that zero venting will be part of normal operating practice. Therefore, this is taken as the base case. Scenarios are then considered for various levels of gas field leakage and venting to illustrate the implications of not applying best practice. Assumptions for LNG production and shipping are as for conventional gas.

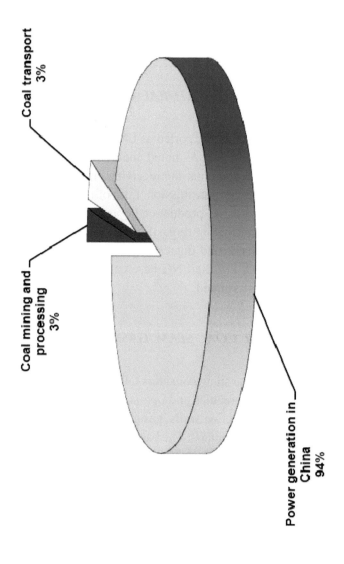

FIGURE 1: Percentage contribution to life cycle GHG emissions: black coal.

TABLE 2: Base case life cycle GHG emissions-black coal.

Activity	GHG emissions intensity				
	Base case (t CO_2-e/t product coal)	%	Sub-critical power generation 33% efficiency (t CO_2-e/MWh)	Super-critical power generation 41% efficiency (t CO_2-e/MWh)	Ultra super-critical power generation 43% efficiency (t CO_2-e/MWh)
MINING					
Mine fugitives	0.0375	1.47	0.0152	0.0122	0.0116
Mine diesel use	0.0114	0.40	0.0046	0.0037	0.0035
Explosives	0.00025	0.01	0.0001	0.0001	0.0001
Slow oxidation	0.00018	0.01	0.0001	0.0001	0.0001
Power consumption	0.0157	0.62	0.0063	0.0051	0.0049
Spontaneous combustion	0.00185	0.07	0.0007	0.0006	0.0006
Scope 3 fuel and electricity	0.0029	0.11	0.0012	0.0009	0.0009
TRANSPORT					
Rail operations	0.00205	0.08	0.0008	0.0007	0.0006
Port handling	0.00161	0.06	0.0007	0.0005	0.0005
Shipping	0.0791	3.11	0.0320	0.0257	0.0245
END USE					
Combustion	2.388	94.02	0.9647	0.7765	0.7403
TOTAL all sources	2.540	100	1.026	0.826	0.788
Range Min			0.75	0.61	0.58
Max			1.56	1.26	1.20

7.3 LIFE CYCLE EMISSIONS

7.3.1 AUSTRALIAN BLACK COAL FOR EXPORT

Australia is the world's largest exporter of black coal. The industry boasts a diversity of mine types (surface, open cut and high wall), sizes, ownership (major and independents), operational conditions and product types. In addition to the existing industry, a large number of new and mine expansion projects are proposed in both New South Wales and Queensland in response to rising prices and world demand for coal, especially from China. GHG emissions sources for each stage of the mining operation are summarized in Table 1.

TABLE 3: Australian conventional LNG: Operations and GHG emissions sources.

Operations	Emissions sources
Extraction and upstream processing • Exploration and test drilling • Gas/water separation, condensate separation, dehydration, compression and other initial processing on offshore platforms • Stripping of CO_2 and other impurities from raw gas • Pipeline transmission to the onshore processing plant	• Operating gas turbines and standby diesel generators power • Flaring or venting gas for safety and during maintenance • Leaks • Emissions from vessels and helicopters • Construction related GHG emissions-transport vessels, diesel generators, helicopters • Embedded emissions in materials and fuel
LNG Facility • Gas treatment to remove impurities, including removal of nitrogen and carbon dioxide • Depending on the plant, some of the gas may be processed for local industrial and domestic use, and transmitted via pipeline • Depending on the plant, processing of condensate for export. Life cycle emission estimates for LNG include apportionment for the export component	• Gas turbines for power generation and liquefaction (largest component of GHG emissions from an LNG plant) • Vented CO_2 from acid gas removal, flared and un-burnt methane from flares and thermal oxidizers • Fugitives from flanges and other leaks (typically small and closely monitored for safety reasons) • Flaring during ship loading (systems are designed to capture boil off gas for use as fuel by the ship) • Construction emissions (diesel generators, plant and vehicles and construction vessels) • Embedded emissions in materials and fuel
Transport The LNG is transported by ship	• Combustion of fuel by the ship • Leaks (for safety reasons leaks from shipboard LNG tanks are typically closely monitored and very small)
Regasification and combustion At or near the destination port the LNG is re-gasified and transmitted by pipeline to the receiving power plant When used for power generation the gas is burned in a combined cycle or open cycle gas turbine plant (base case assumption)	• Energy (gas use) for regasification • Emissions from combustion in the power station

To date, there has been relatively little data available specific to GHG emissions from export of Australian coal. There are extensive project forecast EIS data in Australia, but little publicly available data for existing operations. GHG emissions estimates have been developed from existing information from 6 underground and 9 open cut mines of which some examples are listed in the references [23–27]. The coal mines were selected on the basis of EIS availability, and to reflect a range of mine types, location, and status. These data were combined with existing studies to develop emission estimate ranges.

Base case estimates for GHG life cycle emissions for Australian black coal for export to China are provided in Table 2, broken down by activity. Figure 1 shows the percentage contribution to overall emissions from production, transport and power generation stages of the life-cycle.

The majority of life cycle GHG emissions occur in end use combustion (94%). Extraction and processing in Australia account for only a small component (2.7%). Of extraction and processing activities, fugitive emissions (1.5%) are the largest single contributor, followed by use of fuel and power (1.2%).

7.3.2 CONVENTIONAL LNG FOR EXPORT

Australian conventional natural gas is almost entirely sourced from large offshore wells, complemented by extensive transmission and distribution systems. Much of this infrastructure has been in place for more than a decade. The life cycle GHG emissions of Australian Northwest Shelf conventional gas are already well established. Raw gas composition varies according to location, but typically includes CO_2 and other impurities. GHG emissions sources are summarized in Table 3.

Data for this analysis were drawn from public submissions of EIS documents from a variety of Northwest Shelf LNG projects, and LCA reports based on information from planned and operational plants in Western Australia. Data from the Karratha Gas plant, using gas from the NR2 field (with lower than average CO_2 content in feed gas at around 2%), were used to estimate life cycle emissions as 0.60 and 0.44 t CO_2-e/MWh for OCGT and CCGT respectively, and total emissions intensity of 3.12

t CO_2-e/t LNG [8]. An LCA for the proposed Scarborough LNG project, assuming shipment of LNG to California, included detailed calculation of shipping emissions which have been used in subsequent studies. The average total emissions intensity (including combustion) was estimated at 3.88 t CO_2-e/t LNG (based on 6.3 Mt of LNG delivered) [28]. A recent literature review [29] of LNG liquefaction, transport, and regasification found average emissions intensities 0.006 t CO_2-e per GJ for these stages of the life cycle. Table 4 compares emissions intensities for various existing and proposed liquefaction plants in Australia, and shows that the GHG intensity of LNG depends in part on the CO_2 content of the feed gas. The significant number of proposed LNG projects reflects Australia's emergence as one of the world's major LNG exporters.

TABLE 4: GHG emissions from Western Australia LNG plants (after Barnett, 2010 [29]).

Plant	E/P *	Trains	Inflow CO_2 (mol%)	T CO_2-e/t LNG	G CO_2-e/ MJ
Darwin LNG	E	1	6	0.46	5.17
NWS Karratha	E	5	2.5	0.35	3.76
Gorgon LNG	P	3	14.2 (80% CCS)	0.35	3.97
Wheatstone LNG	P	6	<2	0.37	3.97
Pluto LNG	P	1	1.7	0.32	3.43
Prelude LNG	P	1	NA	0.63	6.76
Ichthys LNG	P	2	17	0.25 (estimate)	8.05
Browse LNG	P	3	12		3.76
Average				0.442	4.89

* E = existing, P = proposed.

Recent US-based studies have found a similar range of intensities. PACE [30] estimated life cycle GHG emissions from imported LNG, accounting for natural gas extraction, liquefaction, shipping, regasification and pipeline transport. The intensity was 0.74 t CO_2-e/t LNG. Jaramillo [31] calculated emissions intensities in the range of 0.69 to 1.68 t CO_2-e/t LNG for the same production and transportation segments.

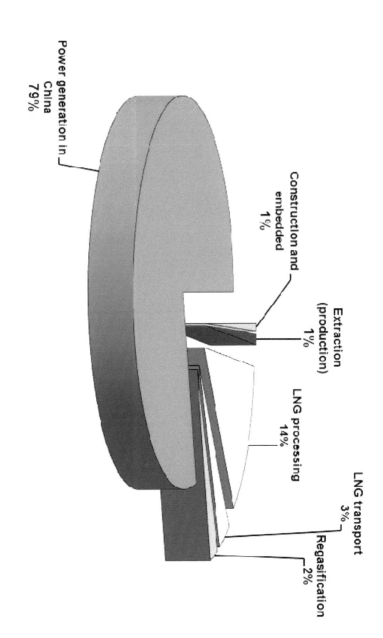

FIGURE 2: Percentage contribution to life cycle GHG emissions: conventional LNG.

FIGURE 3: Breakdown of life cycle GHG emissions from CSG-LNG Reference Case.

Based on the available data, a base-case for GHG emissions for a typical or "average" Australian LNG export project into China is shown in Table 5 for each stage of the life cycle. The base case uses an average Northwest Shelf CO_2 feed gas content, and includes construction and embedded emissions. Combustion in open cycle and combined cycle power-plant scenarios are provided. The ranges in the averages are due mostly to variation in the thermal efficiency of power plants.

TABLE 5: Conventional LNG life cycle GHG emission-base case.

Life cycle operation	Emissions Intensity			
	t CO_2-e/t LNG		t CO_2-e/MWh	
		%	OCGT	CCGT
Assumed average efficiency (%)			41	53
Construction and embedded	0.02-est	0.6	0.003	0.002
Extraction (production)	0.03	0.9	0.005	0.003
LNG processing	0.44	13.6	0.065	0.047
Transport	0.11	3.4	0.03	0.02
Regasification	0.08	2.4	0.02	0.01
Power generation in China	2.54	78.6	0.52	0.36
Totals	3.23	100	0.65	0.45
Ranges Min			0.53	0.39
Max			0.71	0.54

The majority of GHG emissions occur in end-use combustion (79%), but extraction and processing in Australia accounts for a significant component (15%), as shown in Figure 2.

7.3.3 COAL SEAM GAS TO LNG FOR EXPORT

Australian coal seam gas exported from Queensland to China as LNG is used as a reference case. Recent studies have shown that CSG-LNG was less GHG intensive than coal across its life cycle for most end-use combustion scenarios [9,10], based on data from two available EIS reports submitted by project proponents, considering early design proposals and assumed best practice in emissions management, including zero venting and minimal fugitive emissions from leakage (0.1% of production). The CSG industry in Australia is in the early stages of development and data for CSG projects and potential upstream GHG emissions remain limited, largely based on forecasts rather than measured data. The present study considers these GHG emissions in more depth, incorporating more recent information and experience.

TABLE 6: Australian CSG/LNG: Operations and GHG emissions sources.

Operations	Emissions sources
Extraction and upstream processing	• In exploration, use of diesel for drill rigs, and vehicles
• Exploration (including test drilling and core sampling)	
• Drilling of test, pilot, and production wells	• During construction and operation GHG emissions arise from vehicles and machinery, diesel generators, land clearing and embedded emissions in materials and fuel
• Hydraulic fracturing, if required	
• Gas/water separators capture the gas for collection via pipelines to processing plant where the gas is treated (including dehydration) and compressed for transmission	• Flaring and venting from pilot wells, production well completion and work-over
	• Flaring and venting from gas gathering and processing, including compression and dehydration
• High pressure transmission pipeline to the LNG plant	
• Water treatment for reuse or aquifer recharge	• Power for compressors and other systems, including water treatment units
LNG Facility	Similar to conventional gas but raw gas CO2 content is lower
Similar to conventional gas but no condensate production—See Table 3	
Transport	
As for conventional LNG—see Table 3	
Regasification and combustion	
As for conventional gas—see Table 3	

Application of best practice will dictate minimization of fugitive methane emissions. Under a carbon pricing scheme, fugitive methane emissions could lead to significant financial liability for operators. Nevertheless, standard operating practices may require occasional gas venting.

Sources of GHG emissions are summarized in Table 6 (emissions from LNG plant, transport, regasification and combustion operations are identical to those described in Table 4).

TABLE 7: Methane fugitive emissions mitigation measures.

Emissions sources	Mitigation
Venting from pilot wells, well completions and workovers	• Capturing the gas and connecting to supply lines
	• Capturing gas entrained in produced water
	• Flaring where the gas cannot be used
	• Maximizing combustion efficiency of flaring
	• Minimizing time periods for any necessary venting
Venting from compressor stations and pneumatic devices; Some equipment, e.g., pneumatic devices, are specifically designed to vent gas when use in gas systems although it appears that their use will be minimal in Queensland as these devices will run on compressed air.	• Use of grid powered instead of gas powered compressor stations
	• Flaring wherever possible
	• Avoiding cold vents
	• Avoiding pneumatic devices using gas
Leaks	• High integrity equipment
	• Construction, installation and testing to high standards
	• Leak detection programs, including remote sensing
Environmental management	Implementing methane emissions minimization as part of implementing environmental management plans including:
	• Assessment of risks and impacts
	• Objectives, targets, plans and KPIs
	• Training and awareness, including sub-contractors
	• Procedures, including incident management
	• Monitoring
	• Auditing, reporting and corrective action

Fugitive emissions and leaks of methane in CSG production may be unintentional or due to process upsets. The large number of wells required for CSG extraction at scale (between 6000 and 10,000 wells for a large scale CSG development in Queensland), and associated gas handling equipment, pipe work and connections, provide additional potential for GHG emissions. Although fugitives may be a small percentage of total production, the GHG impacts are magnified, since, as noted above, methane's global warming potential is 21 times that of CO_2 over a 100-year period [19], and between 72 to 105 times over a 20-year period. Managing these potential sources of GHG emissions is an important consideration for CSG operators. Upstream fugitive emissions from existing CSG operations are dominated by compressor station venting, field and compression fuel gas consumption, pilot and production well venting, leaks from connections and equipment throughout the gathering system, entrained CH_4 in water production, and system upsets and blowdowns [32]. In estimating fugitive emissions for the CSG-LNG reference case, it is assumed that the current regulatory requirements for fugitive emissions in Queensland are being met, including a "no venting" requirement. A recent government review of 2715 CSG well heads found only five had "reportable" leaks [33]. Avoiding methane venting is already recognized internationally as best practice [34]. However, recent studies from the USA have indicated the potential for significant venting of fugitives if best practice is not followed [12].

There are no current Australian-specific guidelines for estimating natural gas fugitives. Australia's current NGER Technical Guidelines specify using the US API Compendium [17], which may be considered out of date. Emissions factors for equipment used in the US may not be applicable to proposed projects in Australia. The US EPA has, over the past 15 years, monitored fugitives from the US gas industry, and established the Star Program to work with industry on fugitive emissions reduction. The US EPA conducted a major investigation into fugitives from the US natural gas industry in 1997 and found average losses of $1.4 \pm 0.5\%$ from production, transmission and storage [35]. In 2010 it produced an update, announcing upward revisions of these estimates in some cases and new estimates for well completion and work-over (9175 Mcf methane/work-over or completion). The Star program in the USA and similar programs in Canada have shown that methane emissions can be significantly reduced by applying best practice technology and management methods. Some of the main approaches are summarized in Table 7. Unburnt methane from flaring is not expected to be a large source of GHGs as ground flares burn with an efficiency of at least 99.5% and conventional elevated flares burn with an efficiency of 98% [36]. The Australian CSG industry, still in relative infancy, has a golden opportunity to learn from the North American gas experience, and move now to embed best practice in design, construction and operation of CSG projects and associated infrastructure.

An estimate of upstream fugitive emissions for the Queensland reference case was developed based on the most recent available data from operating CSG fields [37]. Projected peak upstream GHG emissions were estimated at 2.8 Mt CO_2-e, assuming 4500 wells required for the 10 Mtpa reference case.

TABLE 8: GHG emissions for CSG-LNG reference case, at maximum production.

Source of emissions	GHGs (t CO_2-e pa)
Core holes-construction	56,600
Pilot Wells-construction + operations	122,100
CSG Fields-construction	278,600
CSG Pipeline- construction	61,300
LNG Plant-construction	173,100
CSG Fields-operations	4,081,000
CSG Pipeline-operations	5000
LNG Plant-operations	3,526,000
LNG shipment	937,000
LNG re-gasification	758,300
Combustion of LNG	30,065,000
Total life cycle emission	Approx. 40,063,000

The base case estimate is based on a typical large coal seam gas development, as described in a number of EIS reports (e.g., [32,37,38]). The base case assumes preparation of 500 core holes for exploration, 300 pilot wells and 6000 production wells. Each production well is assumed to have a lifetime of 15 years, with 1 well completion and 8 workover activities over this lifetime. The development includes a transmission pipeline and

LNG plant capable of producing 10 Mtpa of LNG. The GHG emissions for the CSG-LNG lifecycle base case are shown in Table 8, on the basis that no gas in the flare streams are vented and using a 100-year methane global warming potential.

Table 8 shows that total upstream annual emissions for the reference facility amount to 4.1 Mt CO_2-e, approximately 10% of the total lifecycle GHG emissions of approximately 40 Mt CO_2-e. Upstream fugitive emissions, as defined by the API Compendium (2009) [17], accounted for 0.73 Mt CO_2-e per annum of this total. The largest source of fugitive emissions is from screw and centrifugal compressors. LNG plant operations account for 9% of emissions, with 0.53 Mt CO_2-e per annum arising from fugitive methane emissions. As found in previous studies, end-use combustion emissions overwhelmingly dominate the lifecycle GHG emissions of all types of LNG (Figure 3).

Table 9 shows the GHG emission intensity per tonne of CSG-LNG product and per MWh of power sent out for the base case (0% venting; 100-year methane GWP).

TABLE 9: Base case life cycle GHG emission intensities for CSG-LNG.

Source of emissions	T CO_2-epa/ GJ	T CO_2-e pa/t LNG	%	OCGT 39% efficiency t CO_2-e/MWh	CCGT 53% efficiency t CO_2-e/MWh
Core holes-construction	0.0001	0.006	0.1	0.001	0.001
Pilot Wells-construction + operations	0.0002	0.012	0.3	0.002	0.001
CSG Fields-construction	0.0005	0.028	0.7	0.004	0.003
CSG Pipeline-construction	0.0001	0.006	0.2	0.001	0.001
LNG Plant-construction	0.0003	0.017	0.4	0.003	0.002
CSG Fields-operations	0.0069	0.408	10.2	0.063	0.047
CSG Pipeline-operations	0.00001	0.001	0.01	0.0001	0.0001
LNG Plant-operations	0.0059	0.353	8.8	0.055	0.040
LNG Shipment	0.0016	0.095	2.3	0.015	0.011
LNG Re-gasification	0.0013	0.077	1.9	0.012	0.009
Combustion of LNG	0.0525	3.138	75.0	0.578	0.425
Total	0.069	4.140	100.0	0.733	0.540

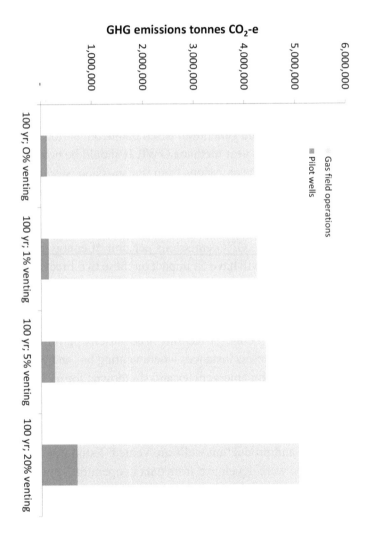

FIGURE 4: Impact of venting scenarios on gas field emissions

The results in Table 9 compare well with other recent lifecycle GHG emissions studies [9] and Jiang et al. [39] for Marcellus shale gas. Jiang et al. [39] considered pre-production, production, processing, transmission, distribution and combustion stages and reported overall lifecycle GHG emissions were 0.068 tonnes/GJ, which is in good agreement with the value of 0.069 tonnes/GJ found in this study. End-use combustion accounted for 75% of lifecycle GHG emissions, with the GHG intensity for electricity sent out from a CCGT power plant ranging from 0.48 to 0.56 t CO2-e/MWh, also in broad agreement with the present study.

The results of the present study also compare well with NETL [40] data for average gas-fired generation based on unconventional gas (0.53 t CO2-e/MWh) for a 100-year methane GWP. It should be noted that Howarth et al. [41] state that the emissions from transmission, storage and distribution reported in Jiang et al. [39] and NETL [40] are 38% and 50% less than those reported by the US EPA [42]. Howarth et al. [41] suggest that this is due to the overestimation of the lifetime gas production from a well, which underestimates the GHG emissions per unit of energy available from gas production. This will have an impact on these two lifecycle GHG emission studies, but the extent of the impacts has not been evaluated here because there is currently very little experience in Australia on the anticipated CSG well life.

Although deliberate gas venting is not strictly permitted in Queensland, there are nevertheless instances where venting has and will continue to occur, such as during emergencies and shutdowns for maintenance. In order to estimate the impact of venting, a number of scenarios were considered in which a percentage of the flare streams were instead vented. Three scenarios were considered, assuming 1%, 5% and 20% of flare streams from pilot wells and production wells are vented. Estimates of annual volumes of gas flared were developed from data in operators' environmental impact statements [37], and included pilot well flaring (2.7 million m^3/year per well), and work-over activities (42,500 m^3/work-over).

Figure 4 shows that GHG emissions during pilot well operations are particularly sensitive to venting of flare stream gas. The base case uses the 100-year methane GWP and 0% venting. A scenario with 1% venting leads to a 24% rise in GHGs from this segment. In terms of CSG field operations, venting of flare streams is less sensitive in terms of overall segment GHG emissions, as these are dominated by combustion of fuel gas in gas engines and compressors. Only a high value of 20% venting leads to a significant change in CSG field GHG emissions (an approximate 7% rise).

In terms of overall lifecycle GHG emissions, only the 20% venting scenario leads to a significant (>2%) change, corresponding to a rise in GHG intensity to 0.55 t CO_2-e/MWh (based on CCGT technology). In the hypothetical situation where all flared gas is vented, the GHG intensity rises to 0.59 t CO_2-e/MWh for a 100-year methane GWP.

A fourth scenario considers the recent results of a sampling campaign in the Denver-Julesberg Fossil Fuel Basin in the United States by Pétron et al. [43]. Various estimates were made of the methane emissions from flashing and venting activities by oil and gas operations in northeastern Colorado. Bottom-up estimates show that 1.68% of the total natural gas produced in 2008 was vented. Top-down scenarios give a range 3.1% up to 4.0% (minimum range of 2.3% up to 3.8% and a maximum range of 4.5% up to 7.7%). In this study we take the average of all top down estimates from Pétron [43] et al., giving 4.38% of all gas production being vented. Although the study of Pétron [43] et al. includes both gas and oil production emissions, no attempt is made here to separate these emission sources. Given that the Denver-Julesberg data represent a field which is several decades old, this clearly represents a worst case scenario when applied to the emerging Australian CSG industry. Nevertheless, it does illustrate what could occur in future if leading practice is not adopted and GHG abatement measures are not incorporated across the industry.

To calculate the impact of the 4.38% loss of CSG as fugitive emissions, the upstream CSG production emissions were also increased commensurately by 4.38% to ensure the same amount of CSG reaches the LNG production facility. This loss of CSG as fugitive emissions results in an additional 8.6 Mt CO_2-e emissions per annum compared with the base case and a 100-year methane GWP. Compared to Figure 3, the emissions from the CSG fields rise from 10% of total lifecycle emissions to 26%, and end-use combustion emissions drop from 75% to 62%. The GHG intensity also rises to 0.64 t CO2-e/MWh for CCGT technology and 0.87 t CO_2-e/MWh for OCGT technology. In this scenario, the lifecycle GHG emissions for OCGT electricity generation are higher than for supercritical and ultra-supercritical coal fired generation.

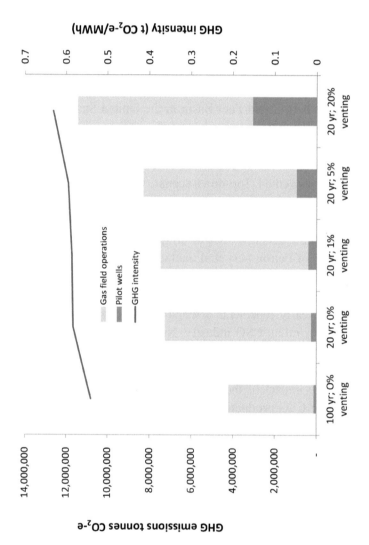

FIGURE 5: Impact of a 20-year methane GWP on upstream GHG emissions and lifecycle emissions intensity for CSG.

Figure 5 shows the impact of changing the methane GWP from the 100-year value of 21 to the 20-year value of 105 [21] on vented emissions. For CSG, the present study finds that the change in methane GWP has an impact on pilot well and gas production well segment emissions. Given the significant volume of gas flared at the pilot well stage (since pilot wells are generally not linked to a gas-gathering pipeline network), any fraction of the gas stream that is vented, instead of being flared, will have an impact on overall GHG emissions. Natural gas venting and leaks from the LNG plant are well-defined and factored into the base case emissions scenario, although a jump in emissions of 0.8 Mt CO_2-e emissions per annum accompanies the increase in methane GWP. For the CSG fields, gas that is vented instead of flared at compressor stations, well completions and workovers, and routine and emergency venting make large contributions to segment GHG emissions. The impact of these releases is amplified by the high 20-year methane GWP.

Figure 5 reflects the large impact of the increase in the methane GWP in pilot well GHGs due to the relatively large amounts of gas flared (of the order of 3 million m^3 of gas is flared per pilot well). For production wells, industry proponents estimate that a total of 25,470 m^3 of CSG are released per well during completions and workovers over a lifetime of 15 years (based on data in [37]: 14,150 m3 flared per day for 3 days during workovers).

In response to the increased methane GWP, the overall lifecycle GHG emissions increase by between 9.6% (3.8 Mt CO_2-e per annum) for 0% flared gas being vented and up to 20% (8 Mt CO_2-e per annum) for 20% of the flare gas being vented. Similarly, the GHG intensity for the CSG/LNG lifecycle rises from 0.54 to 0.63 t CO_2-e/MWh sent out, based on CCGT technology. When the fugitive emissions for coal mining are assessed using the 20-year methane GWP, the GHG intensities also increase, ranging from 0.834 (ultra-supercritical), 0.875 (super-critical) and 1.087 t CO_2-e/MWh (sub-critical). On this basis, the GHG intensity of gas-fired generation is still below the life cycle GHG emissions for all coal-fired generation technologies.

As a comparison, the NETL [40] predicts an intensity of 0.69 t CO_2-e/MWh for average natural gas baseload generation fuelled by shale gas, assuming a 20-year methane GWP of 72. The present study predicts an intensity of 0.63 based on a much higher methane GWP. The variations in the two GHG intensities may be a result of the differences in methane venting volumes for Australian CSG and US shale gas from completions, workovers, and liquid unloading events. Also, gas distribution and storage losses are not a significant part of the Australian CSG/LNG lifecycle as most of the Australian CSG will be converted to LNG for overseas export. Considering the worst case scenario of 4.38% of total upstream production being vented (based on 10 Mtpa of CSG output), and the 20-year methane GWP, results in an additional 41 Mt CO_2-e of emissions per annum. Under this worst case scenario, the GHG intensity of generation using CCGT technology is approximately 1.07 t CO_2-e/MWh sent out, which is higher than ultra-supercritical and super-critical coal-fired generation technology, and nearly the same as sub-critical coal-fired generation when assessed with a 20-year methane GWP.

High losses of CSG through leaks and venting are considered unlikely, as this represents a substantial loss in revenue, a potential safety hazard for the industry, and in Australia, an ongoing significant carbon tax liability. Nevertheless, the results of this analysis indicate the need for the Australian CSG industry to improve monitoring of methane releases and to adopt best practice technology and systems to reduce leaks and venting emissions, particularly during workovers and well completions. Howarth et al. [12] provide a brief review of methane abatement technologies. According to the US General Accountability Office (GAO) [44], "green" technologies are capable of reducing methane emissions by 40%. This includes reducing liquid unloading related emissions with automated plunger lifts and using flash tank separators or vapour recovery units to reduce dehydrator emissions. Reduced emissions completions technologies can reduce emissions from flowbacks during workovers and completions, but this requires gas gathering pipelines to be in place prior to completions. This may not be possible for pilot wells and gas fields under development. Compressor leaks may be reduced by using dry seals and increasing frequency of maintenance and monitoring. Table 7 provides a summary of emissions reduction methods.

From the lifecycle analysis of CSG/LNG, it was apparent that methane releases from liquid unloading, well completion and workover events (whether flared or vented), are potential, yet uncertain, sources of GHG emissions. When compared to the data available in relation to shale gas GHG emissions from the US EPA, it is evident that emissions from these sources require further research in the Australian context. The possibility of methane dissolution and migration in groundwater and subsequent release to atmosphere via improperly abandoned wells or other geological pathways also exists. One study on the Marcellus Shale in the USA found evidence of elevated levels of dissolved methane in groundwater (19.2 mg/L on average), compared to natural background levels (1.1 mg/L), in proximity to gas wells [45]. Given the concentrations reported, the potential for dissolved concentrations of methane in groundwater de-gassing to atmosphere to have a meaningful impact on the overall GHG life-cycle appear small. However, at present, very little research on this migration mechanism and the potential for atmospheric release has been completed, especially in the Australian context.

7.4 LIFE CYCLE GHG EMISSIONS COMPARISON

Using the emissions intensity estimates developed above, GHG emissions of various energy sources were compared in the Australian context for export to China. The base case comparison is between conventional LNG, CSG-LNG and black coal when exported from Australia to China for power generation.

7.4.1 BASE COMPARISON—AUSTRALIAN EXPORT

Table 10 summarizes base case life cycle GHG emissions intensity in electrical power generation in China, for Australian conventional gas, coal seam gas and black coal. Estimates are provided for OCGT and CCGT gas combustion, and for sub-, super-, and ultrasuper-critical coal combustion. The ranges in intensities largely reflect variations in thermal efficiencies in end-use combustion. The base case for CSG/LNG assumes zero venting and leakage losses of 0.1% of production, as discussed above. These findings are provided graphically in Figure 6, including ranges from all life cycle-emissions sources.

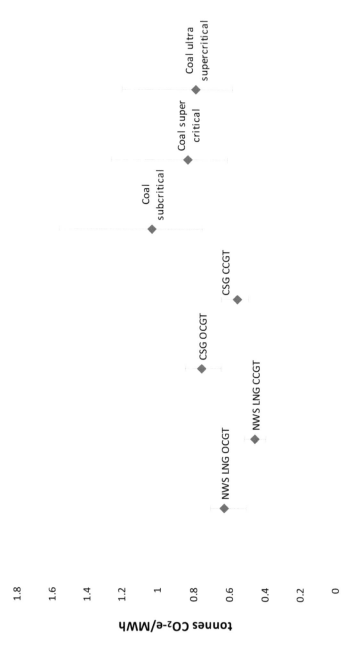

FIGURE 6: Base case GHG intensities and ranges.

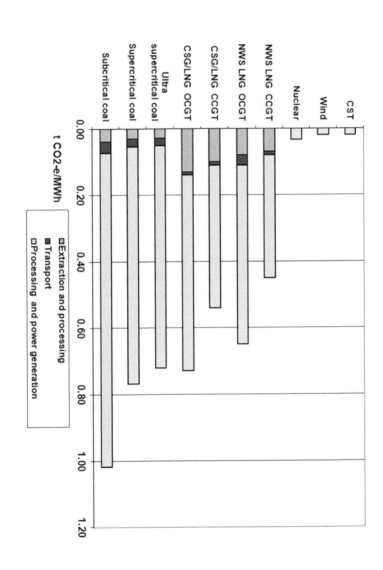

FIGURE 7: Life cycle GHG emissions intensities for Australian fossil fuel exports, and selected renewables and nuclear, base case.

TABLE 10: GHG intensities-base case (t CO_2-e/MWh).

Operation	Conventional gas		Coal seam gas		Black coal		
	OCGT	CCGT	OCGT	CCGT	Subcritical	Super-critical	Ultrasu-percritical
Assumed average efficiency (%)	39	53	39	53	33	41	43
Extraction and processing	0.09	0.07	0.12	0.10	0.03	0.02	0.02
Transport	0.02	0.01	0.02	0.01	0.03	0.03	0.03
Processing and power generation in China	0.54	0.37	0.59	0.43	0.97	0.78	0.74
Totals	0.65	0.45	0.73	0.54	1.03	0.83	0.79
Ranges Min	0.50	0.39	0.64	0.49	0.75	0.61	0.58
Max	0.70	0.51	0.84	0.64	1.56	1.26	1.20
T CO_2-e/t product	3.23	3.23	4.14	4.14	2.54	2.54	2.54

The results show that for all exported fossil fuels, end-use combustion dominates GHG emissions, accounting for 94% of the total in the case of coal, 82% for conventional LNG, and 75% for CSG/LNG. For most combustion technologies, coal is more GHG intensive than LNG. However CSG-LNG is 17–21% more GHG intensive than conventional LNG, largely as a result of higher energy use in upstream production (when zero venting is assumed). Conventional LNG re-gasified and burnt in CCGT power plants is least GHG intensive, and black coal burnt in a subcritical power plant is the most GHG intensive of the scenarios. The gap between coal and LNG narrows considerably with higher efficiency coal technologies and when ranges are considered, to the extent that CSG-LNG burned in low efficiency power plants is slightly more GHG intensive than the most efficient coal combustion technology.

7.4.2 RENEWABLE AND NUCLEAR ENERGY

Renewable and nuclear energy sources provide an alternative basis for comparison of GHG emissions intensities (Table 11). Renewable energy sources (wind, solar, wave and geothermal) produce no GHG emissions in electricity generation, and the GHG intensity is derived from fuel use for construction and ancillary purposes, and embedded emissions in infrastructure and consumables. Wind and concentrated solar thermal (CST) show similar life cycle emissions. Life cycle GHG emissions for nuclear energy depend on the grade of fuel and processing required and how reprocessing power is sourced. Figure 7 illustrates the significantly higher life cycle GHG emissions of exported Australian fossil fuels compared to solar, wind and nuclear when used for power generation in China.

TABLE 11: GHG emission intensities for renewable and nuclear energy-base case.

	Emissions intensity t CO_2-e/MWh	Range (from literature review) t CO_2-e/MWh
Wind	0.021 [13]	0.013–0.040 [13]
Solar Photovoltaic	0.106 [13]	0.053–0.217 [13]
Concentrated Solar Thermal	0.020 [46]	Central tower 0.0202 [46]
		Parabolic trough 0.0196[46]
Hydro	0.015 [13]	0.006–0.044 [13]
Nuclear-current technologies	0.034 [47]	0.01–0.13 [13]

Note: The emissions intensities stated here have been derived from the specific LCA studies referenced and from associated literature reviews of LCA studies conducted internationally. For the purposes of the comparison in this study the figures are applicable to power generation in China.

7.4.3 DISPLACEMENT OF COAL BY GAS

Recent Australian studies have examined the theoretical GHG emissions reductions that could occur if LNG is exported to China and other Asian destinations [8–10]. Depending on the assumptions around generation technology, and assuming full displacement, natural gas exported as LNG was found generally to offer a potential overall global GHG emissions savings. However, the assumption that LNG exported to China, or any other Asian destination, would result in a coal-fired power station being taken off-line and replaced by a gas-fired power station is problematic [9].

The International Energy Agency has recently suggested that while this type of direct displacement is likely in the USA, it is unlikely that LNG will displace coal in Asia. Rather LNG is more likely to add to overall capacity in an expanding energy market [48]. Using the base case estimates from this study, if CCGT combustion technology fueled by natural gas derived from conventional LNG displaced an old subcritical coal-fired power station, 0.58 t CO_2-e/MWh of emissions would be avoided (0.49 t CO_2-e for CSG/LNG). This represents the best average case for displacement by Australian LNG. If natural gas-fired OCGT displaced an ultra-supercritical coal plant, however, the savings would drop to 0.14 and 0.06 t CO_2-e/MWh for conventional and CSG derived LNG, respectively, again assuming base cases.

Currently, coal is relatively cheap compared to gas. However, renewables and nuclear power are more expensive than gas. Under current market conditions, therefore, displacement of renewables by imported LNG in China is also a possible scenario. If LNG-fired conventional OCGT technology were to displace wind or concentrated solar thermal power in China, an overall increase in emissions of 0.63 t CO_2-e/MWh would be experienced, rising to 0.71 t CO_2-e/MWh for CSG/LNG. If global GHG savings are to be claimed as a key driver for LNG development, detailed economic research and modelling should be undertaken to determine the markets and conditions under which real benefits are generated.

7.5 CONCLUSION

This analysis brings together the most recent data available from energy producers and studies available in the literature to produce an average comparison of the lifecycle GHG intensities per MWh of electricity sent out, for a range of Australian and other energy sources. When Australian fossil fuels are exported to China, lifecycle greenhouse gas emission intensity in electricity production depends to a significant degree on the technology used in combustion. In general, natural gas exported as LNG is less GHG intensive than black coal but the gap is smaller for OCGT plant and for CSG.

On average, conventional LNG burned in a conventional OCGT plant is approximately 38% less GHG intensive over its life cycle than black coal burned in a sub-critical plant, per MWh of electricity produced. However, if OCGT combustion is compared to the most efficient new ultra-supercritical coal-fired power, the gap narrows considerably. Coal seam gas LNG is approximately 13–20% more GHG intensive across its life cycle, on a like-for-like basis, than conventional LNG, and thus compares less favorably to coal than conventional LNG under all technology combinations. Upstream fugitive emissions from CSG in the Australian context were found to be uncertain because of a lack of data. Nevertheless, fugitive methane emissions are potentially manageable by applying best practice technologies.

In modelling the GHG emissions for a typical CSG-LNG development, it was assumed that between 1% and 20% of the flare stream gas was vented. Combined with the latest estimate for the 20-year GWP for methane, these vented emissions significantly added to the overall GHG footprint. However, the lifecycle GHG intensity rankings did not materially change, such is the dominance of end-use combustion. The exception to this is if the worst case scenario of 4.38% of all production is released as leaks and vented emissions (based on recent US studies). Here, the GHG intensity of electricity generation using CCGT technology based on CSG/LNG is approximately 1.07 t CO_2-e/MWh sent out, which is higher than ultra-supercritical and super-critical coal-fired generation technology, and nearly the same as sub-critical coal-fired generation when assessed with a 20-year methane GWP.

The implications for regulators and the emerging Australian CSG industry are that best practice applied to design, construction and operation of projects can significantly reduce emissions (particularly fugitives), lower financial liabilities under the carbon tax, and help make CSG a less GHG-intensive fuel option.

When exported for electricity production, LNG was found to be 22 to 36 times more GHG intensive than wind and concentrated solar thermal (CST) power and 13–21 times more GHG intensive than nuclear power. Transitioning the world's energy economy to a lower carbon future will require significant investment in a variety of cleaner technologies, including renewables and nuclear power. In the short term, improving the efficiency of fossil fuel combustion in energy generation can provide an important contribution.

Availability of life cycle GHG intensity data will allow decision-makers to move away from overly simplistic assertions about the relative merits of certain fuels, and focus on the complete picture, especially the critical roles of energy policy, technology selection and application of best practice.

REFERENCES

1. IEA (International Energy Agency). World Energy Outlook; 2010; IEA: Paris, France, November 2010.
2. BP. Statistical Review of World Energy; 2010; BP: London, UK, 2011.
3. BP. BP Energy Outlook; 2010; BP: London, UK, 2011.
4. Hondo, H. Life cycle GHG emission analysis of power generation systems: Japanese case. Energy 2005, 30, 2024–2056.
5. Weisser, D. A guide to life cycle greenhouse gas (GHG) emissions for electricity supply technologies. Energy 2007, 32, 1543–1559.
6. Dones, R.; Heck, T.; Hirschberg, S. Greenhouse gas emissions for energy systems: Comparison and overview. Energy 2004, 27–95.
7. Tamura, I.; Tanaka, T.; Kagajo, T.; Hwabara, S.; Yoshioka, T.; Nagata, T.; Kurahashi, K. Life cycle CO2 analysis of LNG and city gas. Appl. Energy 2001, 68, 301–319.
8. WorleyParsons. Greenhouse Gas Emissions Study of Australian LNG. Available online: http://www.woodside.com.au/Our-Approach/Climate-Change/Pages/Benefits-of-LNG (accessed on 12 September 2011)
9. WorleyParsons. Greenhouse Gas Emissions Study of Australian CSG to LNG; Technical Report for the Association of Petroleum Producers and Explorers of Australia (APPEA); APPEA: Sydney, Australia, 2011.
10. Coal Seam Gas and Greenhouse Emissions-Comparing the Lifecycle Emissions for CSG/LNG vs. Coal; Technical Report for Citi Global Markets; Institute for Sustainable Futures: Sydney, Australia, 2011.
11. US EPA. Global Anthropogenic Non-CO2 Greenhouse Gas Emissions: 1990–2020; US Environmental Protection Agency: Washington, DC, USA, 2006.
12. Howarth, R.W.; Santoro, R.; Ingraffea, A. Methane and the greenhouse-gas footprint of natural gas from shale formations. Clim. Change 2011, 106, 679–690.
13. Lenzen, M. Life cycle energy and greenhouse gas emissions of nuclear energy: A review. Energy Convers. Manag. 2008, 49, 2178–2199.
14. Raadal, H.L.; Gagnon, L.; Modahl, I.S.; Hanssen, O.J. Life cycle greenhouse gas (GHG) emissions from the generation of wind and hydro power. Renew. Sustain. Energy Rev. 2011, 15, 3417–3422.
15. World Business Council for Sustainable Development and the World Resource Institute. The Greenhouse Gas Protocol: A Corporate Accounting and Reporting Stan-

dard; World Business Council for Sustainable Development and the World Resource Institute: Washington, DC, USA, 2004.

16. Australian Government, Department of Climate Change and Energy Efficiency. The National Greenhouse and Energy Reporting (Measurement) Determination; Commonwealth of Australia: Sydney, Australia, 2010.

17. American Petroleum Institute. Compendium of Greenhouse Gas Emissions Estimation Methodologies for the Oil and Natural Gas Industry. Available online: http://www.api.org/ehs/climate/new/upload/2009_GHG_COMPENDIUM.pdf (accessed on 15 March 2009).

18. Australian Government, Department of Climate Change and Energy Efficiency. National Greenhouse and Energy Reporting System Measurement Technical Guidelines; Australian Government: Canberra, Australia, 2010.

19. Intergovernmental Panel on Climate Change. IPCC Second Assessment, Climate Change. Available online: http://www.ipcc.ch/pdf/climate-changes-1995/ipcc-2nd-assessment/2ndassessment-en.pdf (accessed on 15 September 2011).

20. Intergovernmental Panel on Climate Change. IPCC Fourth Assessment Report (AR4), 2007. Working Group 1, The Physical Science Basis. Available online: http://www.ipcc.ch/publications_and_data/ar4/wg1/en/ch2s2-10-2.html. (accessed on 3 November 2011).

21. Shindell, D.T.; Faluvegi, G.; Koch, D.M.; Schmidt, G.A.; Unger, N.; Bauer, S.E. Improved attribution of climate forcing to emission. Science 2009, 326, 716–718.

22. Frischenknect, R.; Althaus, H.-J.; Bauer, C.; Doka, G.; Heck, T.; Jungbluth, N.; Kellenberger, D.; Nemecek, T. The environmental relevance of capital goods in life cycle assessments of products and services. Int. J. LCA 2007, doi:http://dx.doi.org/10.1065/lca2007.02.308

23. Ulan Coal. Ulan Coal Continued Operations Project. Energy and Greenhouse Assessment. Available online: http://www.ulancoal.com.au/EN/OperatingApprovals/Volume%20Six/Appendix14.pdf (accessed on 3 November 2011).

24. Vale Australia. Ellensfield Coal Mine Project, Environmental Impact Statement. Greenhouse Gas Assessment. Available online: http://www.ap.urscorp.com/Projects/EllensfieldEIS/_pdf/Appendix_I2_Greenhouse_Gas_Assessment.pdf (accessed on 3 November 2011).

25. WACJV. Wallarah 2 Coal Project. Air Quality Assessment. Available online: http://www.wallarah.com.au/air.html (accessed on 3 November 2011).

26. Xstrata Coal, Cumnock No 1 Pty. Cumnock Colliery, Environmental Impact Statement. Greenhouse Gas and Energy Assessment. Available online: http://www.xstrataravensworth.com.au/EN/Publications/EAs/Cumnock_WPP_EA_Main_text_DA_123–05–01_M1.pdf (accessed on 3 November 2011).

27. New Hope Coal. New Acland Stage 3 Expansion, Environmental Impact Statement, Chapter 10. Greenhouse gases and Climate Change. Available online: http://www.newhopecoal.com.au/media/6181/chapter10greenhousegasesfinaldone.pdf (accessed on 3 November 2011).

28. Heede, R. LNG Supply Chain Greenhouse Gas Emissions for the Cabrillo Deepwater Port: Natural Gas from Australia to California, Climate Mitigation Services. Available online: http://www.edcnet.org/pdf/Heede_06_LNG_GHG_Anlys.pdf (accessed on 3 November 2011).

29. Barnett, P.J. Life Cycle Assessment of Liquefied Natural Gas and Its Environmental Impact as a Low Carbon Energy Source; University of Southern Queensland: Queensland, Australia, 2010. Available online: http://eprints.usq.edu.au/18409/1/Barnett_2010.pdf (accessed on 3 November 2011).

30. Pace Global Energy Services. Life Cycle Assessment of GHG Emissions from LNG and Coal Fired Generation Scenarios: Assumptions and Results. Available online: http://www.lngfacts.org/resources/LCA_Assumptions_LNG_and_Coal_Feb09.pdf (accessed on 3 November 2011).

31. Jaramillo, P. A Life Cycle Comparison of Coal and Natural Gas for Electricity, Generation and the Production of Transportation Fuels. Ph.D. Thesis, Carnegie Mellon University, Pittsburgh, PA, USA, May 2008. Available online: http://wpweb2.tepper.cmu.edu/ceic/theses/Paulina_Jaramillo_PhD_Thesis.pdf (accessed on 3 November 2011).

32. Arrow Energy Limited. Environmental Management Plan for Authority to Prospect 683, May 2010. Available online: http://www.arrowenergy.com.au/icms_docs/72619_EM_Plan_ATP_683.pdf (accessed on 3 November 2011).

33. Queensland Government, Department of Employment, Economic Development and Innovation. Coal Seam Gas Well Head Safety Program—Final Report, 2011. Available online: http://mines.industry.qld.gov.au/assets/petroleum-pdf/Coal-Seam-Gas-Well-Head-Safety-ProgramInspection-Report-2011.pdf (accessed on 3 November 2011).

34. IEA. World Energy Outlook 2011, The Golden Age of Gas; International Energy Agency: Paris, France, 2011.

35. Harrison, M.R.; Shires, T.M.; Wessels, J.K.; Cowgill, R.M. Methane Emissions from the Natural Gas Industry; Technical Report for Environmental Protection Agency; USEPA: Washington, DC, USA, 1996.

36. Australia Pacific LNG. Environmental Impact Statement, 2010. Available online: http://www.aplng.com.au/our-eis (accessed on 3 November 2011).

37. Queensland Curtis LNG Project Draft and Supplementary EIS, 2010. Available online: http://www.qgc.com.au/01_cms/details.asp?ID=427 (accessed on 3 November 2011).

38. GLNG. GLNG Project Environmental Impact Statement. Available online: http://www.glng.com.au/Content.aspx?p=90 (accessed on 29 February 2011).

39. Jiang, M.; Griffin, W.M.; Hendrickson, C.; Jaramillo, P.; van Briesen, J.; Venkatesh, A. Life cycle greenhouse gas emissions of Marcellus shale gas. Environ. Res. Lett. 2011, 6, doi:10.1088/1748- 9326/6/3/034014.

40. National Energy Technology Laboratory (NETL). Life cycle Greenhouse Gas Analysis of Natural Gas Extraction and Delivery in the United States. Available online: http://www.netl.doe.gov/energy-analyses/refshelf/PubDetails.aspx?Action=View&PubId=386) accessed on 29 February 2011).

41. Howarth, R.W.; Santoro, R.; Ingraffea, A. Venting and leaking of methane from shale gas development: response to Cathles. Clim. Change 2012, DOI 10.1007/s10584–012–0401–0.

42. US EPA Inventory of U.S. Greenhouse Gas Emissions and Sinks: 1990–2009; U.S. Environmental Protection Agency: Washington, DC, USA, April 2011. Available

online: http://epa.gov/climatechange/emissions/usinventoryreport.html (accessed on 13 September 2011)

43. Pétron, G.; Frost, G.; Miller, B.R.; Hirsch, A.I.; Montzka, S.A.; Karion, A.; Trainer, M.; Sweeney, C.; Andrews, A.E.; Miller, L.; Kofler, J.; Bar-Ilan, A.; Dlugokencky, E.J.; Patrick, L.; Moore, J.C.T.; Ryerson, T.B.; Siso, C.; Kolodzey, W.; Lang, P.M.; Conway, T.; Novelli, P.; Masarie, K.; Hall, B.; Guenther, D.; Kitzis, D.; Miller, J.; Welsh, D.; Wolfe, D.; Neff, W.; Tans, P. Hydrocarbon emissions characterization in the Colorado Front Range: A pilot study. J. Geophys. Res. 2012, 117, doi:10.1029/2011JD016360.

44. GAO. Federal oil and gas leases: opportunities exist to capture vented and flared natural gas, which would increase royalty payments and reduce greenhouse gases. GAO-11-34 U.S. General Accountability Office: Washington, DC, USA, October 2010. Available online: http://www.gao.gov/new.items/d1134.pdf (accessed on 13 September 2011)

45. Osborn, S.T.; Vengosh, A.; Warner, N.R.; Jackson, R.B. Methane contamination of drinking water accompanying gas-well drilling and hydraulic fracturing. Proc. Natl. Acad. Sci. USA 2011, 108, 8172–8176.

46. Varun, I.K.B.; Prakash, R. LCA of renewable energy for electricity generation systems—A review. Renew. Sustain. Energy Rev. 2009, 13, 1067–1073.

47. Norgate, T.; Haque, N.; Koltun, P.; Tharumarajah, R. Environmental sustainability of nuclear power: Greenhouse gas status. In Proceedings of Presentation of Study Results to AusIMM International Uranium Conference, Perth, Australia, June 8–9 2011.

48. IEA. World Energy Outlook 2011; International Energy Agency: Paris, France, 2011.

CHAPTER 8

Effects of Energy Production and Consumption on Air Pollution and Global Warming

NNENESI KGABI, CHARLES GRANT, AND JOHANN ANTOINE

8.1 INTRODUCTION

The harvesting, processing, distribution, and use of fuels and other sources of energy have major environmental implications including land-use changes due to fuel cycles such as coal, biomass, and hydropower, which affect both the natural and human environment. Energy systems carry a risk of routine and accidental release of pollutants [1] . Greenhouse gas (GHG) and air pollutant emissions share the same sources—transport, industry, commercial and residential areas [2] . All these sources depend on production, distribution and utilization of energy for their daily activities.

Effects of Energy Production and Consumption on Air Pollution and Global Warming. © Kgabi N, Charles Grant C, and Antoine J. Journal of Power and Energy Engineering 2,8 (2014), DOI:10.4236/ jpee.2014.28003. Licensed under Creative Commons Attribution 4.0 International License, http://creativecommons.org/licenses/by/4.0/.

The gases included in GHG inventories are the direct GHGs: namely, carbon dioxide (CO_2), methane (CH_4), nitrous oxide (N_2O), hydrofluorocarbons (HFCs), perfluorocarbons (PFCs) and sulphur hexafluoride (SF6), and the indirect GHGs: non-methane volatile organic compounds (NMVOC), carbon monoxide (CO), nitrogen oxide (NOx), and sulphur dioxide (SO_2) [2] .

Jamaica has no known primary petroleum or coal reserves and imports all of its petroleum and coal requirements. Domestic energy needs are met by burning petroleum products and coal and renewable fuel biomass (i.e., biogases, fuel wood, and charcoal) and using other renewable resources (e.g., solar, wind and hydro). In 2008, approximately 86 percent of the energy mix was imported petroleum, with the remainder coming from renewables and coal [3]. Electricity is generated primarily by oil-fired steam, engine driven, and gas turbine units. Smaller amounts of electricity are generated by hydroelectric and wind power. Use of solar energy is negligible, and the option for nuclear energy has not been exploited.

The increase in GHG emissions, with the country just a few years from the global warming tipping point is evident. Carbon dioxide emissions increased from 9531 Gg in 2000 to 13,956 Gg in 2005, methane emissions from 31.1 Gg in 2000 to 41.9 Gg in 2005. Nitrous oxide emissions also increased although in smaller quantities. Emissions from the electricity generation source category between 2000 and 2005 ranged from 2977 Gg to 3365 Gg for CO_2, 0.116 Gg to 0.132 Gg for methane, and 0.023 Gg to 0.026 Gg for N_2O [4].

The objective of this study was to assess different fuel combinations that can be adopted to reduce the level of air pollution and GHG emissions associated with the energy. The study bears significance to almost all countries that, due to the pressure of high energy demand, tend to settle for any available energy source without considering the environmental effects.

8.2 METHODS

Desktop study methods were used to source data for this secondary research. Information relating to energy generation and utilization in Jamaica was accessed outside organizational boundaries, mainly from online sources, research journals, professional bodies/organisations and government published data and reports. The data acquisition approach was purposive and mostly "cherry-picking", i.e., based on keyword searches, footnote chases, citation searches or forward chains, journal runs, and to some extent author searches.

Content analysis methods for drawing conclusions included noting patterns, themes and trends, making comparisons, building logical chain of evidence and making conceptual/theoretical coherence. The content analysis yielded some descriptive data giving a detailed picture of the energy generation, air pollution and climate change in Jamaica.

8.3 RESULTS AND DISCUSSION

Electricity generation, transmission, and distribution are associated with GHG emissions like carbon dioxide (CO_2), and smaller amounts of methane (CH_4) and nitrous oxide (N_2O). These gases are released during the combustion of fossil fuels, such as coal, oil, and natural gas, to produce electricity. Less than 1% of greenhouse gas emissions from the electricity sector come from sulfur hexafluoride (SF_6), an insulating chemical used in electricity transmission and distribution equipment [5] .

8.3.1 ELECTRICITY CONSUMPTION

Consumption of electricity has direct GHG emission implications for the company/organization generating the electricity, and indirect implications for the consumer. Table 1 shows the annual increase in GHG emissions with increase in electricity consumption. The main electric utility related gases are: GHGs—CO_2, CH_4, N_2O, SF_6; and Air Pollutants—CO, SO_2, NOx, NMVOCs.

Electricity sales data was obtained from World Bank, Benchmarking data of the electricity distribution sector in Latin America and the Caribbean Region 1995-2005; and the emission factors: CO_2—0.819 tons CO_2/MWh; CH_4—0.03716 kg CO_2/MWh; N_2O—0.00743 ton CO_2/MWh were used for calculation of the GHGs.

8.3.2 ELECTRICITY GENERATION

The annual fuel use for electricity generation between 2000 and 2005 ranged from 5,159,687 to 4,811,726 million barrels of heavy fuel oil and from 725,158 to 1,794,870 million barrels of diesel oil. Global warming potential of the fuel use are summarized in Table2

TABLE 1: Annual electricity sales and GHG emissions (data source for electricity sold [6]; source for emission factors used in the calculations [7]).

	Electricity Sold (MW h)	Methane (tonCO$_2$e)	Nitrous Oxide (tonCO$_2$e)	Carbon Dioxide (tonCO$_2$e)	Total GHG Emissions
2001	2,793,375	103.8	20754.8	2,287,774	2,308,633
2002	2,896,547	107.6	21521.3	2,372,271	2,393,900
2003	2,998,345	111.4	22277.7	2,455,644	2,478,033
2004	2,975,509	110.6	22,108	2,436,942	2,459,161
2005	3,055,154	113.5	22699.8	2,502,171	2,524,984
Average	2,943,786	109.38	21872.32	2410960.4	2,432,942
SD	101,539	3.769	754.431	83160.688	83918.9

Table 2. GHG emissions from combustion of fuel during electricity generation (data source for electricity sold [6] ; source for emission factors used in the calculations [7]).

		Consumption (million barrels)	CO$_2$ ($\times10^6$ tons CO$_2$e)	CH$_4$ ($\times10^6$ ton CO$_2$e)	N$_2$O ($\times10^6$ tons CO$_2$e)	Total GHG ($\times10^6$ tons CO$_2$e)
2000	Fuel Oil	5,159,687	2513	2.16	6.96	2522
	Diesel Oil	725,158	306,322	104	2202	308,627
2005	Fuel Oil	4,811,726	874	0.75	2.42	877
	Diesel Oil	1,794,870	758,190	257	5450	763,897

The emission factors used above were obtained from DEFRA [7] as follows: Diesel = 2.6569, 0.0009, 0.0191, 2.6769 kg CO$_2$e/L; and Fuel Oil = 0.26729, 0.00023, 0.000074, 0.26826 kg CO$_2$e/kWh; for CO$_2$, CH$_4$, N$_2$O, and CO_2e respectively. Other conversions used include 1 barrel = 158.99 liters; and 1 kWh = 0.00009 tonne oil equivalent.

Choice of the right fuel mix for electricity generation determines the amount of air pollutants and GHGs released into the atmosphere. The current electricity fuel mix of Jamaica is fuel oil (71%), diesel oil (24%) and 5% renewable. Figure 1 shows the fuel mix used in 2007 based on the installed capacity by energy sources (MWh) data obtained from the United States Energy Information Administration (EIA) [8] .

8.3.3 THE ENERGY FUEL MIX

Jamaica's National Energy Policy 2009-2030: contribution of fuel mix to electricity generation mix is summarized in Figure 2.

The 3.3 percent increase in annual electricity generation (GWh) over the period (1998-2009), moving from 2950 GWh in 1998 to 4214 GWh in 2009; was used as baseline to estimate possible implications of the National Energy Policy on GHG emissions as shown in Figure 3.

Contribution by each fuel type is shown in Figure 4, emphasizing the importance of a correct fuel mix in reduction of GHG emissions.

Conversions used include: Natural Gas—CO_2 = 0.18483, CH_4 = 0.00027, N_2O = 0.00011, CO_2e = 0.18521 kg COe/unit; Coal-CO_2 = 0.32360, CH_4 = 0.00006, N_2O = 0.00282, CO_2e = 0.32648 kg COe/unit; and LPG-CO_2 = 0.21419, CH_4 = 0.00010, N_2O = 0.00025, CO_2e = 0.21455 kg COe/unit.

8.3.4 ELECTRICITY DISTRIBUTION

Figure 5 shows annual percentages of losses that occur during distribution of electricity in Jamaica. The data used was obtained from the United States Energy Information Administration (EIA). Electrical transmissions and distribution systems contribute significantly to emissions of sulfur hexafluoride (SF_6), which is also a GHG. The losses during distribution also add to the emissions.

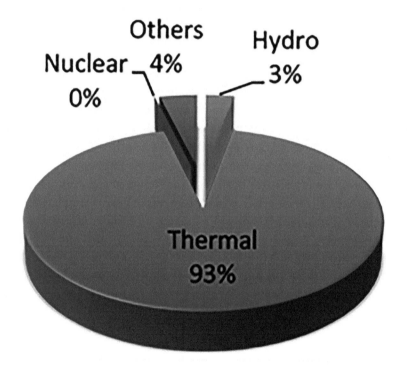

FIGURE 1: 2007 electricity generation fuel mix (data source: [8]).

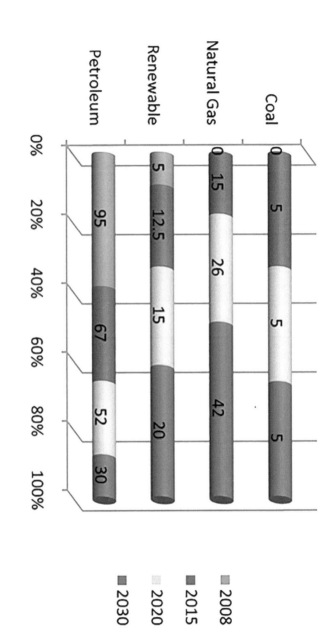

FIGURE 2: Electricity generation fuel mix proposed in the National Energy Policy (data source: [9]).

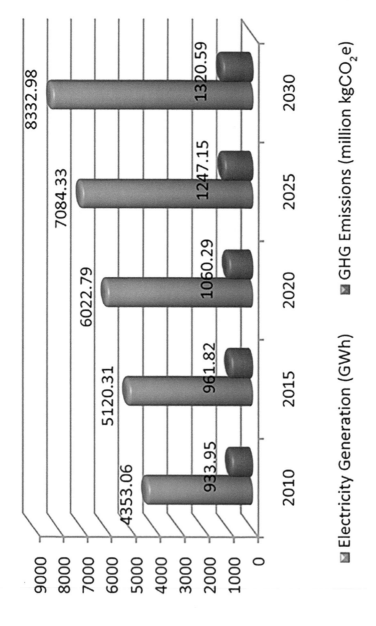

FIGURE 3: Estimate energy production and the potential GHG emissions (data source: [9]).

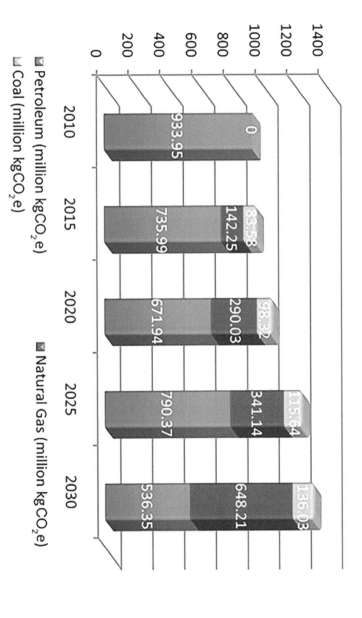

FIGURE 4: Contribution of the proposed fuels to GHG emissions (data source: [2] , [7]).

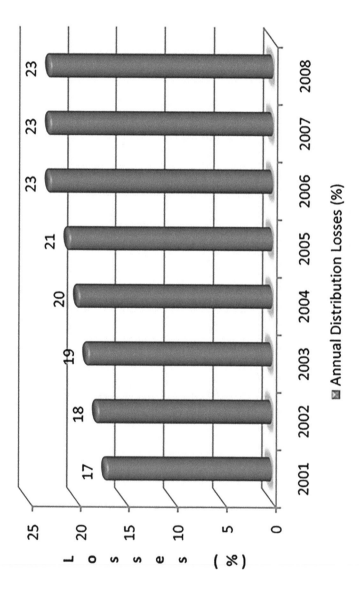

FIGURE 5: Electricity distribution losses (data source: [8]).

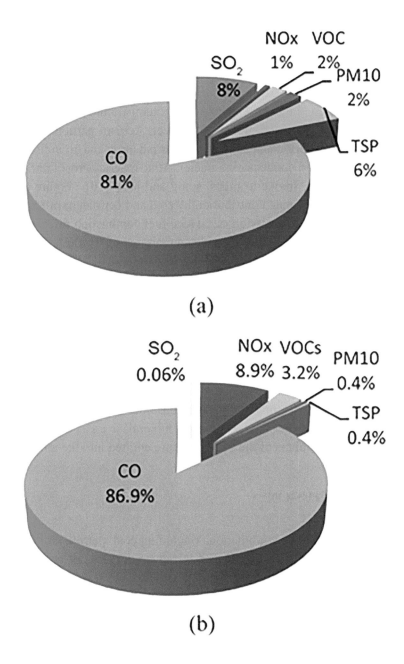

FIGURE 6: (a) Coal air pollutants; (b) Petroleum air pollutants (data source: [11]).

8.3.5 AIR QUALITY IMPLICATIONS

Trace gases and aerosols impact climate through their effect on the radiative balance of the earth. Trace gases such as greenhouse gases absorb and emit infrared radiation which raises the temperature of the earth's surface causing the enhanced greenhouse effect. Aerosol particles have a direct effect by scattering and absorbing solar radiation and an indirect effect by acting as cloud condensation nuclei. Atmospheric aerosol particles range from dust and smoke to mists, smog and haze [10] . Figure 6 gives the average air pollutant contribution by coal and petroleum products.

In addition to emission of GHGs, fuel combustion affects air quality. Combustion of 1 kg of coal results in emission of 19 g SO_2, 1.5 g NOx, 5 g VOCs, 4.1 g PM10, 14.7 g TSP, 187.4 g CO and 0.0134 g benzene; while 1 kg of petroleum products emits 0.01 g SO_2, 1.4 g NOx, 0.5 g VOCs, 0.07 g PM10, 0.07 g TSP, and 13.6 g CO into the atmosphere [11].

On combustion, fuel oil also produces primarily carbon dioxide and water vapour, but also smaller quantities of particulate matter and oxides of nitrogen and sulphur (OUR, 2012).

The potential environmental problem associated with coal is the formation of acidic effluents due to pyrite oxidation and the consequent mobilization of environmentally hazardous trace elements, which are mainly associated with the sulphide group of minerals in coal [12] . With combustion, these amounts of the SOx gases are emitted into the atmosphere.

8.4 CONCLUSIONS

The different fuel combinations (including coal, petroleum products, and natural gas) that can be adopted to reduce the level of air pollution and GHG emissions associated with the energy were assessed. This study has shown that: 1) choice of the fuel mix determines the success of GHG emissions reductions; and 2) there is no single fuel that is not associated with GHG or other air pollution or environmental degradation implications. The usual increase in GHG emissions with increase in energy consumption and production was observed.

Given the increasing energy demand and the environmental implications of the fuel mix options discussed, it may be necessary to also explore the nuclear energy option. Nuclear power plants do not require a lot of space when compared to equivalent wind or solar farms. The nuclear energy does not contribute to carbon emissions (no CO_2 is given out) thus does not cause global warming. Production and consumption of the nuclear energy do not produce smoke particles to pollute the atmosphere. The great advantage of nuclear power is its enormous energy density, several million times that of chemical fuels. Even without recycling, one kilogram of oil produces about 4 kWh; a kg of uranium fuel generates 400,000 kWh of electricity. This also reduces transport costs (although the fuel is radioactive and therefore each transport that does occur is expensive because of security implications). Furthermore, it produces a small volume of waste.

REFERENCES

1. Holdren, J.P. and Smith, K.R. (2000) Energy, the Environment, and Health. In: World Energy Assessment: Energy and the Challenge of Sustainability, 63-110.
2. IPCC (2006) Guidelines for National Greenhouse Gas Inventories.http://www.ipcc-nggip.iges.or.jp/public/2006gl/index.html
3. Office of Utilities Regulation (OUR) (2012) Sulphur Content of Fuel Oil Used for Power Generation.www.cwjamaica.com/-office.our
4. Jamaica Productivity Centre (2010) Generation and Distribution of Electricity in Jamaica: A Regional Comparison of Performance Indicators.
5. United States Environmental Protection Agency (USEPA) (2012) Sources of Greenhouse Gas Emissions.http://www.epa.gov/climatechange/ghgemissions/sources/electricity.html
6. World Bank (2013) Benchmarking Data of the Electricity Distribution Sector in Latin America & the Caribbean Region 1995-2005. http://info.worldbank.org/etools/lacelectricity/home.htm
7. DEFRA (2012) 2012 Guidelines to DEFRA/DECC's GHG Conversion Factors for Company Reporting. AEA for the Department of Energy and Climate Change (DECC) and the Department for Environment, Food and Rural Affairs (DEFRA).
8. United States Energy Information Administration (EIA) (2012). http://www.iea.org/stats/index.asp
9. Ministry of Energy and Mining (MEM) (2009) Jamaica's National Energy Policy 2009-2030.
10. IPCC (2001) Climate Change 2001: The Scientific Basis, Contribution of Working Group I to the Third Assessment Report of the Intergovernmental Panel on Climate Change.

11. Friedl, A., et al. (2004) Air Pollution in Dense Low-Income Settlements in South Africa. Royal Danish Embassy, Department of Environmental Affairs and Tourism, 2008.

12. Garcia, A.B. and Martinez-Tarazona, M.R. (1993) Removal of Trace Elements from Spanish Coals by Flotation. Fuel, 72, 329-335. http://dx.doi.org/10.1016/0016-2361(93)90050-C

CHAPTER 9

Venting and Leaking of Methane from Shale Gas Development: Response to Cathles et al.

ROBERT W. HOWARTH, RENEE SANTORO, AND ANTHONY INGRAFFEA

9.1 INTRODUCTION

Promoters view shale gas as a bridge fuel that allows continued reliance on fossil fuels while reducing greenhouse gas (GHG) emissions. Our April 2011 paper in Climatic Change challenged this view (Howarth et al. 2011). In the first comprehensive analysis of the GHG emissions from shale gas, we concluded that methane emissions lead to a large GHG footprint, particularly at decadal time scales. Cathles et al. (2012) challenged our work. Here, we respond to the criticisms of Cathles et al. (2012), and show that most have little merit. Further, we compare and contrast our assumptions and approach with other studies and with new information made available since our paper was published. After carefully considering all of these, we stand by the analysis and conclusions we published in Howarth et al. (2011).

9.2 METHANE EMISSIONS DURING ENTIRE LIFE CYCLE FOR SHALE GAS AND CONVENTIONAL GAS

Cathles et al. (2012) state our methane emissions are too high and are "at odds with previous studies." We strongly disagree. Table 1 compares our estimates for both conventional gas and shale gas (Howarth et al. 2011) with 9 other studies, including 7 that have only become available since our paper was published in April 2011, listed chronologically by time of publication. See Electronic Supplementary Materials for details on conversions and calculations. Prior to our study, published estimates existed only for conventional gas. As we discussed in Howarth et al. (2011), the estimate of Hayhoe et al. (2002) is very close to our mean value for conventional gas, while the estimate from Jamarillo et al. (2007) is lower and should probably be considered too low because of their reliance on emission factors from a 1996 EPA report (Harrison et al. 1996). Increasing evidence over the past 15 years has suggested the 1996 factors were low (Howarth et al. 2011). In November 2010, EPA (2010) released parts of their first re-assessment of the 1996 methane emission factors, increasing some emissions factors by orders of magnitude. EPA (2011a), released just after our paper was published in April, used these new factors to re-assess and update the U.S. national GHG inventory, leading to a 2-fold increase in total methane emissions from the natural gas industry.

The new estimate for methane emissions from conventional gas in the EPA (2011a) inventory, 0.38 g C MJ^{-1}, is within the range of our estimates: 0.26 to 0.96 g C MJ^{-1} (Table 1). As discussed below, we believe the new EPA estimate may still be too low, due to a low estimate for emissions during gas transmission, storage, and distribution. Several of the other recent estimates for conventional gas are very close to the new EPA estimate (Fulton et al. 2011; Hultman et al. 2011; Burnham et al. 2011). The Skone et al. (2011) value is 29% lower than the EPA estimate and is very similar to our lower-end number. Cathles et al. (2012) present a range of values, with their high end estimate of 0.36 g C MJ^{-1} being similar to the EPA estimate but their low end estimate (0.14 g C MJ^{-1}) far lower than any other estimate, except for the Jamarillo et al. (2007) estimate based on the old 1996 EPA emission factors.

TABLE 1: Comparison of published estimates for full life-cycle methane emissions from conventional gas and shale gas, expressed per unit of Lower Heating Value (gC MJ−1). Studies are listed by chronology of publication date

	Conventional gas	Shale gas
Hayhoe et al. (2002)	0.57	*
Jamarillo et al. (2007)	0.15	*
Howarth et al. (2011)	0.26–0.96	0.55–1.2
EPA (2011a)	0.38	0.60+
Jiang et al. (2011)	*	0.30
Fulton et al.(2011)	0.38++	*
Hultman et al. (2011)	0.35	0.57
Skone et al. (2011)	0.27	0.37
Burnham et al. (2011)	0.39	0.29
Cathles et al. (2012)	0.14–0.36	0.14–0.36

See Electronic Supplemental Materials for details on conversions
*Estimates not provided in these reports
+Includes emissions from coal-bed methane, and therefore may under-estimate shale gas emissions
++Based on average for all gas production in the US, not just conventional gas, and so somewhat over-estimates conventional gas emissions

For shale gas, the estimate derived from EPA (2011a) of 0.60 g C MJ^{-1} is within our estimated range of 0.55 to 1.2 g C MJ^{-1} (Table 1); as with conventional gas, we feel the EPA estimate may not adequately reflect methane emissions from transmission, storage, and distribution. Hultman et al. (2011) provide an estimate only slightly less than the EPA number. In contrast, several other studies present shale gas emission estimates that are 38% (Skone et al. 2011) to 50% lower (Jiang et al. 2011; Burnham et al. 2011) than the EPA estimate. The Cathles et al. (2012) emission estimates

are 40% to 77% lower than the EPA values, and represent the lowest estimates given in any study.

In an analysis of a PowerPoint presentation by Skone that provided
the basis for Skone et al. (2011), Hughes (2011a) concludes that a major
difference between our work and that of Skone and colleagues was the
estimated lifetime gas production from a well, an important factor since
emissions are normalized to production. Hughes (2011a) suggests that
Skone significantly overestimated this lifetime production, and thereby
underestimated the emissions per unit of energy available from gas production (see Electronic Supplemental Materials). We agree, and believe
this criticism also applies to Jiang et al. (2011). The lifetime production
of shale-gas wells remains uncertain, since the shale-gas technology is so
new (Howarth and Ingraffea 2011). Some industry sources estimate a 30-
year lifetime, but the oldest shale-gas wells from high-volume hydraulic
fracturing are only a decade old, and production of shale-gas wells falls
off much more rapidly than for conventional gas wells. Further, increasing
evidence suggests that shale-gas production often has been exaggerated
(Berman 2010; Hughes 2011a, 2011b; Urbina 2011a, 2011b).

Our high-end methane estimates for both conventional gas and shale
gas are substantially higher than EPA (2011a) (Table 1), due to higher
emission estimates for gas storage, transmission, and distribution ("downstream" emissions). Note that our estimated range for emissions at the
shale-gas wells ("upstream" emissions of 0.34 to 0.58 g C MJ^{-1}) agree very
well with the EPA estimate (0.43 g C MJ^{-1}; see Electronic Supplementary
Materials). While EPA has updated many emission factors for natural gas
systems since 2010 (EPA 2010, 2011a, 2011b), they continue to rely on
the 1996 EPA study for downstream emissions. Updates to this assumption currently are under consideration (EPA 2011a). In the meanwhile, we
believe the EPA estimates are too low (Howarth et al. 2011). Note that the
downstream emission estimates of Hultman et al. (2011) are similar to
EPA (2011a), while those of Jiang et al. (2011) are 43% less, Skone et al.
(2011) 38% less, and Burnham et al. (2011) 31% less (Electronic Supplemental Materials). One problem with the 1996 emission factors is that they
were not based on random sampling or a comprehensive assessment of
actual industry practices, but rather only analyzed emissions from model
facilities run by companies that voluntarily participated (Kirchgessner et

al. 1997). The average long-distance gas transmission pipeline in the U.S. is more than 50 years old, and many cities rely on gas distribution systems that are 80 to 100 years old, but these older systems were not part of the 1996 EPA assessment. Our range of estimates for methane emissions during gas storage, transmission, and distribution falls well within the range given by Hayhoe et al. (2002), and our mean estimate is virtually identical to their "best estimate" (Howarth et al. 2011). Nonetheless, we readily admit that these estimates are highly uncertain. There is an urgent need for better measurement of methane fluxes from all parts of the natural gas industry, but particularly during completion of unconventional wells and from storage, transmission, and distribution sectors (Howarth et al. 2011).

EPA proposed new regulations in October 2009 that would require regular reporting on GHG emissions, including methane, from natural gas systems (EPA 2011c). Chesapeake Energy Corporation, the American Gas Association, and others filed legal challenges to these regulations (Nelson 2011). Nonetheless, final implementation of the regulations seems likely. As of November 2011, EPA has extended the deadline for the first reporting to September 2012 (EPA 2011c). These regulations should help evaluate methane pollution, although actual measurements of venting and leakage rates will not be required, and the reporting requirement as proposed could be met using EPA emission factors. Field measurements across a range of well types, pipeline and storage systems, and geographic locations are important for better characterizing methane emissions.

9.3 HOW MUCH METHANE IS VENTED DURING COMPLETION OF SHALE-GAS WELLS?

During the weeks following hydraulic fracturing, frac-return liquids flow back to the surface, accompanied by large volumes of natural gas. We estimated substantial methane venting to the atmosphere at this time, leading to a higher GHG footprint for shale gas than for conventional gas (Howarth et al. 2011). Cathles et al. (2012) claim we are wrong and assert that methane emissions from shale-gas and conventional gas wells should be equivalent. They provide four arguments: 1) a physical argument that large flows of gas are not possible while frac fluids fill the well; 2) an assertion that venting of methane to the atmosphere would be unsafe; 3) a statement that we incorrectly used data on methane capture during flowback to estimate venting; and 4) an assertion that venting of methane is not in the economic interests of industry. We disagree with each point, and note our methane emission estimates during well completion and flowback are quite consistent with both those of EPA (2010, 2011a, b) and Hultman et al. (2011).

Cathles et al. state that gas venting during flowback is low, since the liquids in the well interfere with the free flow of gas, and imply that this condition continues until the well goes into production. While it is true that liquids can restrict gas flow early in the flow-back period, gas is freely vented in the latter stages. According to EPA (2011d), during well cleanup following hydraulic fracturing "backflow emissions are a result of free gas being produced by the well during well cleanup event, when the well also happens to be producing liquids (mostly water) and sand. The high rate backflow, with intermittent slugs of water and sand along with free gas, is typically directed to an impoundment or vessels until the well is fully cleaned up, where the free gas vents to the atmosphere while the water and sand remain in the impoundment or vessels." The methane emissions are "vented as the backflow enters the impoundment or vessels" (EPA 2011d). Initial flowback is 100% liquid, but this quickly becomes a two-phase flow of liquid and gas as backpressure within the fractures declines (Soliman & Hunt 1985; Willberg et al. 1998; Yang et al. 2010; EPA 2011a, d). The gas produced is not in solution, but rather is free-flowing with the liquid in this frothy mix. The gas cannot be put into production and sent to sales until flowback rates are sufficiently decreased to impose pipeline pressure.

Is it unsafe for industry to vent gas during flowback, as Cathles et al. assert? Perhaps, but venting appears to be common industry practice, and the latest estimates from EPA (2011b, page 3–12) are that 85% of flowback gas from unconventional wells is vented and less than 15% flared or captured. While visiting Cornell, a Shell engineer stated Shell never flares gas during well completion in its Pennsylvania Marcellus operations (Bill Langin, pers. comm.). Venting of flow-back methane is clearly not as unsafe as Cathles et al. (2012) believe, since methane has a density that is only 58% that of air and so would be expected to be extremely buoyant when vented. Under sufficiently high wind conditions, vented gas may be mixed and advected laterally rather than rising buoyantly, but we can envision no atmospheric conditions under which methane would sink into a layer over the ground. Buoyantly rising methane is clearly seen in Forward Looking Infra Red (FLIR) video of a Pennsylvania well during flowback (Fig. 1). Note that we are not using this video information to infer any information on the rate of venting, but simply to illustrate that venting occurred in the summer of 2011 in Pennsylvania and that the gas rose rapidly into the atmosphere. Despite the assertion by Cathles et al. that venting is illegal in Pennyslvania, the only legal restriction is that "excess gas encountered during drilling, completion or stimulation shall be flared, captured, or diverted away from the drilling rig in a manner than does not create a hazard to the public health or safety" (PA § 78.73. General provision for well construction and operation).

Cathles et al. state with regard to our paper: "The data they cite to support their contention that fugitive methane emissions from unconventional gas production is [sic] significantly greater than that from conventional gas production are actually estimates of gas emissions that were captured for sale. The authors implicitly assume that capture (or even flaring) is rare, and that the gas captured in the references they cite is normally vented directly into the atmosphere." We did indeed use data on captured gas as a surrogate for vented emissions, similar to such interpretation by EPA (2010). Although most flowback gas appears to be vented and not captured (EPA 2011b), we are aware of no data on the rate of venting, and industry apparently does not usually measure or estimate the gas that is vented during flowback. Our assumption (and that of EPA 2010) is that the rate of gas flow is the same during flowback, whether vented or captured. Most of the data we used were reported to the EPA as part of their "green completions" program, and they provide some of the very few publicly available quantitative estimates of methane flows at the time of flowback. Note that the estimates we published in Howarth et al. (2011) for emissions at the time of well completion for shale gas could be reduced by 15%, to account for the estimated average percentage of gas that is not vented but rather is flared or captured and sold (EPA 2011b). Given the other uncertainty in these estimates, though, our conclusions would remain the same.

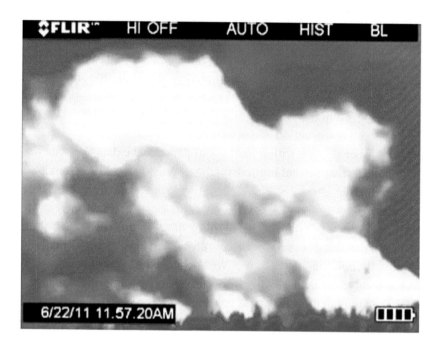

FIGURE 1: Venting of natural gas into the atmosphere at the time of well completion and flowback following hydraulic fracturing of a well in Susquehanna County, PA, on June 22, 2011. Note that this gas is being vented, not flared or burned, and the color of the image is to enhance the IR image of this methane-tuned FLIR imagery. The full video of this event is available at http://www.psehealthyenergy.org/resources/view/198782. Video provided courtesy of Frank Finan

Cathles et al. also assert that we used initial production rates for gas wells, and that in doing so over-estimated flowback venting. Our estimates of flowback emissions for the Barnett, Piceance, Uinta, and Denver-Jules basins were not based on initial production rates, but rather solely on industry-reported volumes of gas captured, assuming. We estimated emissions for the Haynesville basin as the median of data given in Eckhardt et al. (2009), who reported daily rates ranging from 400,000 m³ (14 MMcf) to 960,000 m3 (38 MMcf). We assumed a 10-day period for the latter part of the flowback in which gases freely flow, the mean for the other basin studies we used. The use of initial production rates applied to the latter portion of flowback duration as an estimate of venting is commonly accepted (Jiang et al. 2011; NYS DEC 2011).

Finally, Cathles et al. state that economic self-interest would make venting of gas unlikely. Rather, they assert industry would capture the gas and sell it to market. According to EPA (2011b), the break-even price at which the cost of capturing flowback gas equals the market value of the captured gas is slightly under $4 per thousand cubic feet. This is roughly the well-head price of gas over the past two years, suggesting that indeed industry would turn a profit by capturing the gas, albeit a small one. Nonetheless, EPA (2011b) states that industry is not commonly capturing the gas, probably because the rate of economic return on investment for doing so is much lower than the normal expectation for the industry. That is, industry is more likely to use their funds for more profitable ventures than capturing and selling vented gas (EPA 2011b). There also is substantial uncertainty in the cost of capturing the gas. At least for low-energy wells, a BP presentation put the cost of "green" cleanouts as 30% higher than for normal well completions (Smith 2008). The value of the captured gas would roughly pay for the process, according to BP, at the price of gas as of 2008, or approximately $6.50 per thousand cubic feet (EIA 2011a). At this cost, industry would lose money by capturing and selling gas not only at the current price of gas but also at the price forecast for the next 2 decades (EPA 2011b).

In July 2011, EPA (2011b, e) proposed new regulations to reduce emissions during flowback. The proposed regulation is aimed at reducing ozone and other local air pollution, but would also reduce methane emissions. EPA (2011b, e) estimates the regulation would reduce flowback methane emissions from shale gas wells by up to 95%, although gas capture would only be required for wells where collector pipelines are already in place, which is often not the case when new sites are developed. Nonetheless, this is a very important step, and if the regulation is adopted and can be adequately enforced, will reduce greatly the difference in emissions between shale gas and conventional gas in the U.S. We urge universal adoption of gas-capture policies.

To summarize, most studies conclude that methane emissions from shale gas are far higher than from conventional gas: approximately 40% higher, according to Skone et al. (2011) and using the mean values from Howarth et al. (2011), and approximately 60% higher using the estimates from EPA (2011a) and Hultman et al. (2011). Cathles et al. assertion that shale gas emissions are no higher seems implausible to us. The suggestion by Burnham et al. (2011) that shale gas methane emissions are less than for conventional gas seems even less plausible (see Electronic Supplementary Materials).

9.4 TIME FRAME AND GLOBAL WARMING POTENTIAL OF METHANE

Methane is a far more powerful GHG than carbon dioxide, although the residence time for methane in the atmosphere is much shorter. Consequently, the time frame for comparing methane and carbon dioxide is critical. In Howarth et al. (2011), we equally presented two time frames, the 20 and 100 years integrated time after emission, using the global warming potential (GWP) approach. Note that GWPs for methane have only been estimated at time scales of 20, 100, and 500 years, and so GHG analyses that compare methane and carbon dioxide on other time scales require a more complicated atmospheric modeling approach, such as that used by Hayhoe et al. (2002) and Wigley (2011). The GWP approach we follow is quite commonly used in GHG lifecycle analyses, sometimes considering both 20-year and 100-year time frames as we did (Lelieveld et al. 2005; Hultman et al. 2011), but quite commonly using only the 100-year time frame (Jamarillo et al. 2007; Jiang et al. 2011; Fulton et al. 2011; Skone et al. 2011; Burnham et al. 2011). Cathles et al. state that a comparison based on the 20-year GWP is inappropriate, and criticize us for having done so. We very strongly disagree.

Considering methane's global-warming effects at the decadal time scale is critical (Fig. 2). Hansen et al. (2007) stressed the need for immediate control of methane to avoid critical tipping points in the Earth's climate system, particularly since methane release from permafrost becomes increasingly likely as global temperature exceeds 1.8°C above the baseline average temperature between 1890 and 1910 (Hansen and Sato 2004; Hansen et al. 2007). This could lead to a rapidly accelerating positive feedback of further global warming (Zimov et al. 2006; Walter et al. 2007). Shindell et al. (2012) and a recent United Nations study both conclude that this 1.8°C threshold may be reached within 30 years unless societies take urgent action to reduce the emissions of methane and other short-lived greenhouse gases now (UNEP/WMO 2011). The reports predict that the lower bound for the danger zone for a temperature increase leading to climate tipping points—a 1.5°C increase—will occur within the next 18 years or even less if emissions of methane and other short-lived radiatively active substances such as black carbon are not better controlled, beginning immediately (Fig. 2) (Shindell et al. 2012; UNEP/WMO 2011).

In addition to different time frames, studies have used a variety of GWP values. We used values of 105 and 33 for the 20- and 100-year integrated time frames, respectively (Howarth et al. 2011), based on the latest information on methane interactions with other radiatively active materials in the atmosphere (Shindell et al. 2009). Surprisingly, EPA (2011a) uses a value of 21 based on IPCC (1995) rather than higher values from more recent science (IPCC 2007; Shindell et al. 2009). Jiang et al. (2011), Fulton et al. (2011), Skone et al. (2011), and Burnham et al. (2011) all used the 100-year GWP value of 25 from IPCC (2007), which underestimates methane's warming at the century time scale by 33% compared to the more recent GWP value of 33 from Shindell et al. (2009). We stand by our use of the higher GWP values published by Shindell et al. (2009), believing it appropriate to use the best and most recent science. While there are considerable uncertainties in GWP estimates, inclusion of the suppression of photosynthetic carbon uptake due to methane-induced ozone (Sitch et al. 2007) would further increase methane's GWP over all the values discussed here.

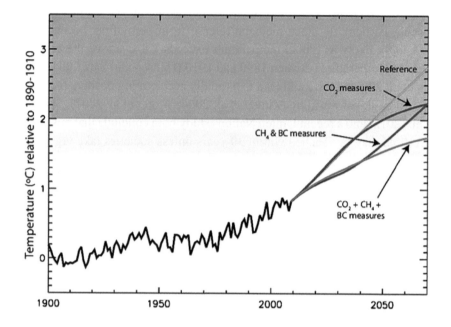

FIGURE 2: Observed global mean temperature from 1900 to 2009 and projected future temperature under four scenarios, relative to the mean temperature from 1890–1910. The scenarios include the IPCC (2007) reference, reducing carbon dioxide emissions but not other greenhouse gases ("CO_2 measures"), controlling methane and black carbon emissions but not carbon dioxide ("CH_4 + BC measures"), and reducing emissions of carbon dioxide, methane, and black carbon ("CO_2 + CH_4 + BC measures"). An increase in the temperature to 1.5° to 2.0°C above the 1890–1910 baseline (illustrated by the yellow bar) poses high risk of passing a tipping point and moving the Earth into an alternate state for the climate system. The lower bound of this danger zone, 1.5° warming, is predicted to occur by 2030 unless stringent controls on methane and black carbon emissions are initiated immediately. Controlling methane and black carbon shows more immediate results than controlling carbon dioxide emissions, although controlling all greenhouse gas emissions is essential to keeping the planet in a safe operating space for humanity. Reprinted from UNEP/WMO (2011)

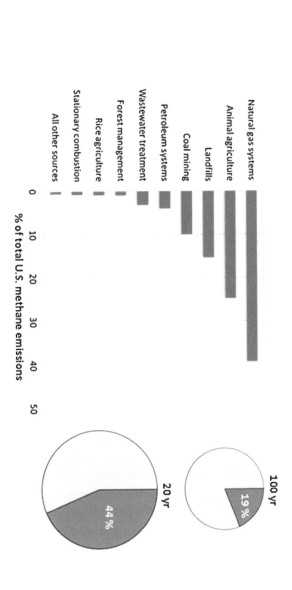

FIGURE 3: Environmental Protection Agency estimates for human-controlled sources of methane emission from the U.S. in 2009 (bar graph and percent contribution of methane to the entire greenhouse gas inventory for the U.S. (shown in red on the pie charts) for the 100-year and 20-year integrated time scales. The sizes of the pie charts are proportional to the total greenhouse gas emission for the U.S. in 2009. The methane emissions represent a greater portion of the warming potential when converted to equivalents of mass of carbon dioxide at the shorter time scale, which increases both the magnitude of the total warming potential and the percentage attributed to methane. Data are from EPA (2011a, b), as discussed in Electronic Supplemental Material, and reflect an increase over the April 2011 national inventory estimates due to new information on methane emissions from Marcellus shale gas and tight-sand gas production for 2009 (EPA 2011b). Animal agriculture estimate combines enteric fermentation with manure management. Coal mining combines active mines and abandoned mines. The time-frame comparisons are made using the most recent data on global warming potentials from Shindell et al. (2009)

In Fig. 3, we present the importance of methane to the total GHG inventory for the US, considered at both the 20- and 100-year time periods, and using the Shindell et al. (2009) GWP values. Figure 3 uses the most recently available information on methane fluxes for the 2009 base year, reflecting the new methane emission factors and updates through July 2011 (EPA 2010; 2011a, b); see Electronic Supplemental Materials. Natural gas systems dominate the methane flux for the US, according to these EPA estimates, contributing 39% of the nation's total. And methane contributes 19% of the entire GHG inventory of the US at the century time scale and 44% at the 20-year scale, including all gases and all human activities. The methane emissions from natural gas systems make up 17% of the entire anthropogenic GHG inventory of the US, when viewed through the lens of the 20-year integrated time frame. If our high-end estimate for downstream methane emissions during gas storage, transmission, and distribution is correct (Howarth et al. 2011), the importance of methane from natural gas systems would be even greater.

9.5 ELECTRICITY VS. OTHER USES

Howarth et al. (2011) focused on the GHG footprint of shale gas and other fuels normalized to heat from the fuels, following Lelieveld et al. (2005) for conventional gas. We noted that for electricity generation—as opposed to other uses of natural gas—the greater efficiency for gas shifts the comparison somewhat, towards the footprint of gas being less unfavorable. Nonetheless, we concluded shale gas has a larger GHG footprint than coal even when used to generate electricity, at the 20-year time horizon (Howarth et al. 2011). Hughes (2011b) further explored the use of shale gas for electricity generation, and supported our conclusion. Cathles et al. criticize us for not focusing exclusively on electricity.

We stand by our focus on GHG emissions normalized to heat content. Only 30% of natural gas in the U.S. is used to generate electricity, while most is used for heat for domestic, commercial, and industrial needs, and this pattern is predicted to hold over coming decades (EIA 2011b; Hughes 2011b). Globally, demand for heat is the largest use of energy, at 47% of use (International Energy Agency 2011). And natural gas is the largest source of heat globally, providing over half of all heat needs in developed countries (International Energy Agency 2011). While generating electricity from natural gas has some efficiency gains over using coal, we are aware of no such advantage for natural gas over other fossil fuels for providing heat.

Many view use of natural gas for transportation as an important part of an energy future. The "Natural Gas Act" (H.R.1380) introduced in Congress in 2011 with bipartisan support and the support of President Obama would provide tax subsidies to encourage long-distance trucks to switch from diesel to natural gas (Weiss and Boss 2011). And in Quebec, industry claims converting trucks from diesel to shale gas could reduce GHG emissions by 25 to 30% (Beaudine 2010). Our study suggests this claim is wrong and indicates shale gas has a larger GHG footprint than diesel oil, particularly over the 20-year time frame (Howarth et al. 2011). In fact, using natural gas for long-distance trucks may be worse than our analysis suggested, since it would likely depend on liquefied natural gas, LNG. GHG emissions from LNG are far higher than for non-liquified gas (Jamarillo et al. 2007). See Electronic Supplemental Materials for more information on future use of natural gas in the U.S.

9.6 CONCLUSIONS

We stand by our conclusions in Howarth et al. (2011) and see nothing in Cathles et al. and other reports since April 2011 that would fundamentally change our analyses. Our methane emission estimates compare well with EPA (2011a), although our high-end estimates for emissions from downstream sources (storage, transmission, distribution) are higher. Our estimates also agree well with earlier papers for conventional gas (Hayhoe et al. 2002; Lelieveld et al. 2005), including downstream emissions. Several other analyses published since April of 2011 have presented significantly lower emissions than EPA estimates for shale gas, including Cathles et al. but also Jiang et al. (2011), Skone et al. (2011), and Burnham et al. (2011). We believe these other estimates are too low, in part due to over-estimation of the lifetime production of shale-gas wells.

We reiterate that all methane emission estimates, including ours, are highly uncertain. As we concluded in Howarth et al. (2011), "the uncertainty in the magnitude of fugitive emissions is large. Given the importance of methane in global warming, these emissions deserve far greater study than has occurred in the past. We urge both more direct measurements and refined accounting to better quantify lost and unaccounted for gas." The new GHG reporting requirements by EPA will provide better information, but much more is needed. Governments should encourage and fund independent measurements of methane venting and leakage. The paucity of such independent information is shocking, given the global significance of methane emissions and the potential scale of shale gas development.

We stress the importance of methane emissions on decadal time scales, and not focusing exclusively on the century scale. The need for controlling methane is simply too urgent, if society is to avoid tipping points in the planetary climate system (Hansen et al. 2007; UNEP/WMO 2011; Shindell et al. 2012). Our analysis shows shale gas to have a much larger GHG footprint than conventional natural gas, oil, or coal when used to generate heat and viewed over the time scale of 20 years (Howarth et al. 2011). This is true even using our low-end methane emission estimates, which are somewhat lower than the new EPA (2011a) values and comparable to those of Hultman et al. (2011). At this 20-year time scale, the emissions data from EPA (2011a, b) show methane makes up 44% of the entire GHG inventory for the U.S., and methane from natural gas systems make up 17% of the entire GHG inventory (39% of the methane component of the inventory).

We also stress the need to analyze the shale-gas GHG footprint for all major uses of natural gas, and not focus on the generation of electricity alone. Of the reports published since our study, only Hughes (2011b) seriously considered heat as well as electricity. Cathles et al. (2012), Jiang et al. (2011), Fulton et al. (2011), Hultman et al. (2011), Skone et al. (2011), and Wigley (2011) all focus just on the generation of electricity. We find this surprising, since only 30% of natural gas in the U.S. is used to generate electricity. Other uses such as transportation should not be undertaken without fully understanding the consequences on GHG emissions, and none of the electricity-based studies provide an adequate basis for such evaluation.

Can shale-gas methane emissions be reduced? Clearly yes, and proposed EPA regulations to require capture of gas at the time of well completions are an important step. Regulations are necessary to accomplish emission reductions, as economic considerations alone have not driven such reductions (EPA 2011b). And it may be extremely expensive to reduce leakage associated with aging infrastructure, particularly distribution pipelines in cities but also long-distance transmission pipelines, which are on average more than 50 years old in the U.S. Should society invest massive capital in such improvements for a bridge fuel that is to be used for only 20 to 30 years, or would the capital be better spent on constructing a smart electric grid and other technologies that move towards a truly green energy future?

We believe the preponderance of evidence indicates shale gas has a larger GHG footprint than conventional gas, considered over any time scale. The GHG footprint of shale gas also exceeds that of oil or coal when considered at decadal time scales, no matter how the gas is used (Howarth et al. 2011; Hughes 2011a, b; Wigley et al. 2011). Considered over the century scale, and when used to generate electricity, many studies conclude that shale gas has a smaller GHG footprint than coal (Wigley 2011; Hughes 2011b; Hultman et al. 2011), although some of these studies biased their result by using a low estimate for GWP and/or low estimates for methane emission (Jiang et al. 2011; Skone et al. 2011; Burnham et al. 2011). However, the GHG footprint of shale gas is similar to that of oil or coal at the century time scale, when used for other than electricity generation. We stand by the conclusion of Howarth et al. (2011): "The large GHG footprint of shale gas undercuts the logic of its use as a bridging fuel over coming decades, if the goal is to reduce global warming."

REFERENCES

1. Beaudine M (2010) In depth: shale gas exploration in Quebec. The Gazette, November 15, 2010
2. Berman A (2010) Shale gas—Abundance or mirage? Why the Marcellus shale will disappoint expectations. The Oil Drum, Drumbeat, October 29, 2010. http://www.theoildrum.com/node/7079

3. Burnham A, Han J, Clark CE, Wang M, Dunn JB, and Rivera IP (2011) Life-cycle greenhouse gas emissions of shale gas, natural gas, coal, and petroleum. Environ Sci Technol. doi:10.1021/es201942m
4. Cathles LM, Brown L, Taam M, Hunter A (2012)
5. Eckhardt M, Knowles B, Maker E, Stork P (2009) IHS U.S Industry Highlights. (IHS) Houston TX. Feb-Mar 2009. http://www.gecionline.com/2009-prt-7-final-reviews
6. EIA (2011a) U.S. Natural Gas Wellhead Price (Dollars per Thousand Cubic Feet). U.S. Department of Energy, Energy Information Agency http://www.eia.gov/dnav/ng/hist/n9190us3m.htm
7. EIA (2011b) Annual Energy Outlook 2011. U.S. Department of Energy, Energy Information Agency (released April 2011)
8. EPA (2010) Greenhouse Gas Emissions Reporting from the Petroleum and Natural Gas Industry. Background Technical Support Document. U.S. Environmental Protection Agency, Washington DC. http://www.epa.gov/climatechange/emissions/downloads10/Subpart-W_TSD.pdf
9. EPA (2011a) Inventory of U.S. Greenhouse Gas Emissions and Sinks: 1990–2009. April 14, 2011. U.S. Environmental Protection Agency, Washington DC. http://epa.gov/climatechange/emissions/usinventoryreport.html
10. EPA (2011b) Regulatory Impact Analysis: Proposed New Source Performance Standards and Amendments to the National Emissions Standards for Hazardous Air Pollutants for the Oil and Gas Industry. July 2011. U.S. Environmental Protection Agency, Office of Air and Radiation. Washington DC
11. EPA (2011c) Climate Change – Regulatory Initiatives. U.S. Environmental Protection Agency, Washington DC. http://www.epa.gov/climatechange/emissions/ghgrulemaking.html (downloaded Nov. 22, 2011)
12. EPA (2011d) Oil and natural gas sector: standards of performance for crude oil and natural gas production, transmission, and distribution. EPA-453/R-11-002. Prepared for U.S. EPA Office of Air Quality Planning and Standards by EC/R Incorporated. U.S. Environmental Protection Agency, Washington DC
13. EPA (2011e) Proposed Amendments to Air Regulations for the Oil and Gas Industry Fact Sheet. U.S. Environmental Protection Agency, Washington DC. http://www.epa.gov/airquality/oilandgas/pdfs/20110728factsheet.pdf
14. Fulton M, Mellquist N, Kitasei S, and Bluestein J (2011) Comparing greenhouse gas emissions from natural gas and coal. 25 Aug 2011. Worldwatch Institute/Deutsche Bank
15. Hansen J, Sato M (2004) Greenhouse gas growth rates. Proc Natl Acad Sci USA 101:16 109–16 114
16. Hansen J, Sato M, Kharecha P, Russell G, Lea DW, Siddall M (2007) Climate change and trace gases. Phil Trans R Soc A 365:1925–1954
17. Harrison MR, Shires TM, Wessels JK, Cowgill RM (1996) Methane emissions from the natural gas industry. Volume 1: executive summary. EPA-600/R-96-080a. U.S. Environmental Protection Agency, Office of Research and Development, Washington, DC

18. Hayhoe K, Kheshgi HS, Jain AK, Wuebbles DJ (2002) Substitution of natural gas for coal: climatic effects of utility sector emissions. Clim Chang 54:107–139

19. Howarth RW, Ingraffea A (2011) Should fracking stop? Yes, it is too high risk. Nature 477:271–273

20. Howarth RW, Santoro R, Ingraffea A (2011) Methane and the greenhouse gas footprint of natural gas from shale formations. Climatic Chang Lett. doi:10.1007/s10584-011-0061-5

21. Hughes D (2011a) Lifecycle greenhouse gas emissions from shale gas compared to coal: an analysis of two conflicting studies. Post Carbon Institute, Santa Rosa, http://www.postcarbon.org/reports/PCI-Hughes-NETL-Cornell-Comparison.pdf

22. Hughes D (2011b) Will Natural Gas Fuel America in the 21st Century? Post Carbon Institute, Santa Rosa, CA. http://www.postcarbon.org/report/331901-will-natural-gas-fuel-america-in

23. Hultman N, Rebois D, Scholten M, Ramig C (2011) The greenhouse impact of unconventional gas for electricity generation. Environ Res Lett 6:044008. doi:10.1088/1748-9326/6/4/044008

24. International Energy Agency (2011) Cogeneration and renewables: solutions for a low-carbon energy future. International Energy Agency, Paris

25. IPCC (1995) IPCC Second Assessment, Climate Change, 1995. http://www.ipcc.ch/pdf/climate-changes-1995/ipcc-2nd-assessment/2nd-assessment-en.pdf

26. IPCC (2007) IPCC Fourth Assessment Report (AR4), Working Group 1, The Physical Science Basis. http://www.ipcc.ch/publications_and_data/ar4/wg1/en/contents.html

27. Jamarillo P, Griffin WM, Mathews HS (2007) Comparative life-cycle air emissions of coal, domestic natural gas, LNG, and SNG for electricity generation. Environ Sci Technol 41:6290–6296

28. Jiang M, Griffin WM, Hendrickson C, Jaramillo P, vanBriesen J, Benkatesh A (2011) Life cycle greenhouse gas emissions of Marcellus shale gas. Environ Res Lett 6:034014. doi:10.1088/1748-9326/6/3/034014

29. Kirchgessner DA, Lott RA, Cowgill RM, Harrison MR, Shires TM (1997) Estimate of methane emissions from the US natural gas industry. Chemosphere 35:1365–1390

30. Lelieveld J, Lechtenbohmer S, Assonov SS, Brenninkmeijer CAM, Dinest C, Fischedick M, Hanke T (2005) Low methane leakage from gas pipelines. Nature 434:841–842

31. Nelson G (2011) Natural gas, electronics groups sue EPA. Energy Tribune. Feb. 2, 2011. http://www.energytribune.com/articles.cfm/6480/Natural-Gas-Electronics-Groups-Sue-EPA

32. NYS DEC (2011) SGEIS on the Oil, Gas and Solution Mining Regulatory Program: Well Permit Issuance for Horizontal Drilling and High-Volume Hydraulic Fracturing to Develop the Marcellus Shale and Other Low-Permeability Gas Reservoirs. Revised draft, Sept 2011. New York State Dept. of Environmental Conservation, Albany, NY

33. Shindell DT, Faluvegi G, Koch DM, Schmidt GA, Unger N, Bauer SE (2009) Improved attribution of climate forcing to emissions. Science 326:716–718

34. Shindell D, Kuylenstierna JCI, Vignati E, van Dingenen R, Amann M, Klimont Z, Anenberg SC, Muller N, Janssens-Maenhout G, Raes F, Schwartz J, Faluvegi G, Pozzoli L, Kupiainen K, Höglund-Isaksson L, Emberson L, Streets D, Ramanathan V, Hicks K, Oanh NTK, Milly G, Williams M, Demkine V, Fowler D (2012) Simultaneously mitigating near-term climate change and improving human health and food security. Science 335:183–189. doi:10.1126/science.1210026

35. Sitch S, Cox PM, Collins WJ, Huntingford C (2007) Indirect radiative forcing of climate change through ozone effects on the land-carbon sink. Nature 448:791–794

36. Skone TJ, Littlefield J, Marriott J (2011) Life cycle greenhouse gas inventory of natural gas extraction, delivery and electricity production. Final report 24 Oct 2011 (DOE/NETL-2011/1522). U.S. Department of Energy, National Energy Technology Laboratory, Pittsburgh, PA

37. Smith GR (2008) Reduced Emission (Green) Completion in Low Energy Reserves. BP America Production Company. 15th Annual Natural Gas STAR Implementation Workshop, San Antonio TX, Nov. 11–13, 2008. http://www.epa.gov/gasstar/workshops/annualimplementation/2008.html

38. Soliman MY Hunt JL (1985) Effect of fracturing fluid and its cleanup on well performance. SPE 14514. Presented paper. SPE Regional Meeting, Morgantown WV, 6–8 Nov 1985

39. UNEP/WMO (2011) Integrated assessment of black carbon and tropospheric ozone: summary for decision makers. United Nations Environment Programme and the World Meteorological Organization, Nairobi

40. Urbina I (2011a) Insiders sound an alarm amid a natural gas rush. New York Times, June 26, 2011. http://www.nytimes.com/2011/06/26/us/26gas.html?_r=1&ref=ianurbina

41. Urbina I (2011b) Behind veneer, doubts on natural gas. New York Times, June 26, 2011. http://www.nytimes.com/2011/06/27/us/27gas.html?ref=ianurbina

42. Walter KM, Smith LC, Chapin FS (2007) Methane bubbling from northern lakes: present and future contributions to the methane budget. Phil Trans R Soc A 365:1657–1676

43. Weis DJ, Boss S (2011) Conservatives Power Big Oil, Stall Cleaner Natural Gas Vehicles. Center for American Progress, June 6, 2011. http://www.americanprogress.org/issues/2011/06/nat_gas_statements.html

44. Wigley TML (2011) Coal to gas: the influence of methane leakage. Climatic Chang Lett. doi:10.1007/s10584-011-0217-3

45. Willberg DM, Steinsberger N, Hoover R, Card RJ, Queen J (1998) Optimization of fracture cleanup using flowback analysis. SPE 39920. Presented paper. SPE Rocky Mountain Regional/ Low-permeability Reservoirs Symposium and Exhibition, Denver CO, 5–8 April 1998

46. Yang JY, Holditch SA, McVay DA (2010) Modeling fracture-fluid cleanup in tight-gas wells. Paper SPE 119624, SPE Journal 15(3)

47. Zimov SA, Schuur EAG, Chapin FS (2006) Permafrost and the global carbon budget. Science 312:1612–1613

There are several supplemental files that are not available in this version of the article. To view this additional information, please use the citation on the first page of this chapter.

PART V

RESIDENTIAL EMISSIONS

CHAPTER 10

Life-Cycle Greenhouse Gas Inventory Analysis of Household Waste Management and Food Waste Reduction Activities in Kyoto, Japan

TAKESHI MATSUDA, JUNYA YANO, YASUHIRO HIRAI, AND SHIN-ICHI SAKAI

10.1 INTRODUCTION

Recycling of biodegradable waste has attracted much interest from governments and researchers worldwide as a means of reducing greenhouse gas (GHG) emissions. A number of countries around the world (e.g., South Korea, European countries) have incorporated biodegradable waste recycling into their waste management systems (European Commission 2008; Kim and Kim 2010). In Japan, 220 out of 1800 Japanese local governments implement separate collection of household food waste, although incinera-

Life-cycle Greenhouse Gas Inventory Analysis of Household Waste Management and Food Waste Reduction Activities in Kyoto, Japan. © Matsuda T, Yano J, Hirai Y, and Sakai S-I. The International Journal of Life Cycle Assessment *17,6 (2012). doi: 10.1007/s11367-012-0400-4. Licensed under a Creative Commons Attribution License, http://creativecommons.org/licenses/by/3.0/.*

tion is still a common practice for food waste disposal (MOE, Japan 2008). Several studies have included life-cycle assessment and cost–benefit analysis of biodegradable waste recycling schemes (Eriksson et al. 2005; European Commission 2010; Fukushima et al. 2008; Sonesson et al. 2000; Sakai et al. 2005; Inaba et al. 2010). Sonesson et al. (2000) developed the ORWARE software to evaluate the environmental burdens associated with waste treatment processes such as incineration, composting, and anaerobic digestion. Eriksson et al. (2005) compared the environmental impacts of several waste management scenarios using ORWARE and concluded that anaerobic digestion reduces more GHG emissions than other treatment methods such as incineration and controlled landfilling. These findings are consistent with the results of other studies (European Commission 2010; Fukushima et al. 2008; Sakai et al. 2005; Inaba et al. 2010).

There are two methods to separate biodegradable waste for biological treatment: source-separated collection and nonseparated collection followed by mechanical sorting [commonly referred to as mechanical biological treatment (MBT)]. MBT has been developed and used in European countries (European Commission 2008). Although there are no MBT facilities in Japan, similar technologies have been developed to separate food and paper waste from mixed household waste (Asano et al. 2005; Takuma 2005; Tatara et al. 2010). Whereas nonseparated collection coupled with mechanical sorting has little impact on household waste disposal behavior, source-separated collection has been reported to reduce food waste by changing the household behavior (WRAP 2009; European Commission 2010). The decision-making model proposed by Hirose (1995) can be used to explain such a change in behavior to a form that is more environmentally friendly. The presence of source-separated collection encourages waste generators to visualize the amount of food waste for each household, which in turn stimulates the perception of seriousness, perception of responsibility, and evaluation of feasibility, thereby promoting food waste reduction (Fig. 1). The prevention of edible food waste, termed food loss, is important from the viewpoint of its associated reduced environmental impact (WRAP 2009; Davis and Sonesson 2008; Cuellar and Webber 2010).

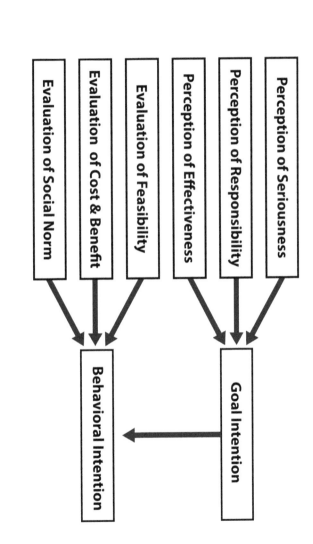

FIGURE 1: Decision-making model of environmentally friendly behavior (Hirose 1995)

In general, waste reduction leads to reductions of both the environmental burden and the amount of recyclables. If recycling is very effective and the life-cycle environmental impact of waste is negative, then the reduction in waste may be environmentally undesirable. Conversely, if recycling of waste is not effective, then waste reduction will be environmentally desirable. Thus, the benefit of waste reduction and the cost of reduced recyclables must be quantified to evaluate the total outcome of waste reduction measures.

Matsuda et al. (2010a, b) studied this tradeoff between anaerobic digestion and the prevention of food loss and concluded that even if anaerobic digestion were implemented, the reduction in GHG as a result of decreased food production would outweigh the reduction of recyclable food waste. The European Commission (2010) assumed that the introduction of separate collection of biowaste would decrease household biowaste by 7.5% as of 2020, mainly via food loss prevention; therefore, the commission analyzed the environmental and financial impact of widespread separate collection.

The results of this study indicated that GHG reduction via food loss prevention in European Union countries could be three or four times higher than the reduction in response to separate collection without waste prevention. However, these two studies only considered food loss prevention, whereas other activities, such as water draining and home composting, can also reduce food waste. Accordingly, these latter activities should be taken into account to better understand the impact of separate collection of food waste.

In this study, we conducted a life-cycle inventory (LCI) analysis of household waste management by incineration and anaerobic digestion. Regional composting is not discussed here because it is not feasible in metropolitan areas like Kyoto due to the lack of local demand for the compost. Our LCI analysis compared two methods for separating food and paper waste from mixed household waste: source-separated collection and nonseparated collection followed by mechanical sorting. Finally, we analyzed three food waste reduction activities: food loss prevention, water draining, and home composting.

TABLE 1: Household waste[a] generation in Kyoto,[b] FY 2008 (from April 2008 to March 2009)

Waste categories	Subcategories	No reduction	Waste reduction cases[c]		
			PrevLoss	ReducDrain	ReducHcom
Food	Cooking waste	42,369	42,369	36,702	36,702
	Leftovers	17,305	10,932	14,990	14,990
	Untouched food	12,087	7,636	10,471	10,471
	Tea leaves and coffee residues	9,168	9,168	7,942	7,942
Paper	Recyclable papers	17,040	17,040	17,040	17,040
	Paper packaging	13,960	13,960	13,960	13,960
	Disposable diapers	6,702	6,702	6,702	6,702
	Other papers	29,379	29,379	29,379	29,379
Plastics		28,300	28,300	28,300	28,300
Wood and grasses[d]		9,447	9,447	9,447	9,447
Other		30,733	30,733	30,733	30,733
Total		216,490	205,666	205,666	205,666
Component (%wet)	Water	41.3	39.7	38.2	39.6
	VS	49.1	50.2	51.7	50.5
	Ash	9.6	10.0	10.1	9.9

[a]Data are expressed as tons wet waste/year
[b]In 2008, the population of Kyoto was 1,464,990
[c]Data in bold font indicates waste reduction
[d]Wood and grasses are wastes composed of wood or grasses (e.g., garden waste, packaging and containers made of wood, furniture made of wood)

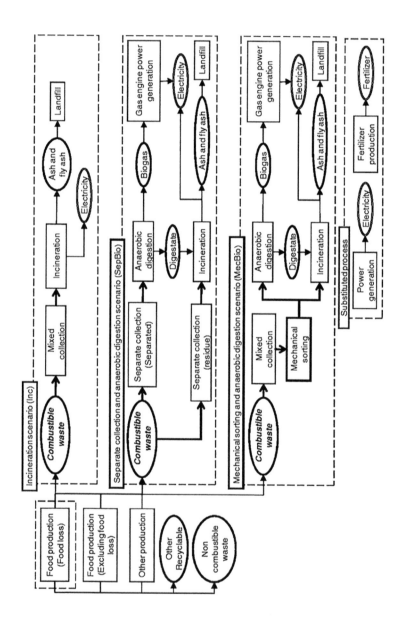

FIGURE 2: System boundary and flow diagram

10.2 METHODS

10.2.1 FUNCTIONAL UNITS

Cleary (2010) proposed an approach known as waste management and prevention life-cycle assessment (LCA; WasteMAP LCA) to evaluate the environmental impact of recycling as well as waste prevention. Following this approach, we set two functional units: a primary functional unit for waste management and a secondary functional unit for waste prevention. The primary functional unit of our study was the annual management of household combustible waste in Kyoto, Japan. Although some of the LCA scenarios included food waste reduction measures, all of the scenarios had an identical secondary functional unit, the annual food ingestion (mass and composition) by the residents of Kyoto, Japan.

The amount of household combustible waste is shown in Table 1 (Matsuda et al. 2010b; Kyoto City Environmental Policy Bureau 2009a, b). Here, combustible waste does not include directly landfilled waste or already recycled waste, such as newspapers. To best represent differences in collection rates for recycling, we divided paper waste into four categories: recyclable paper, paper packaging, disposable diapers, and other papers. Newspaper, cardboard, and similar papers can be recycled back to paper; therefore, we refer to them collectively as recyclable paper for the purposes of this discussion. Other papers such as used tissue paper are especially difficult to recycle. In this study, only unrecyclable types of paper such as disposable diapers and other papers are the source separation targets.

10.2.2 SCENARIO SETTING

We studied the three waste management scenarios shown in Fig. 2 to evaluate the effects of recycling. The incineration scenario (Inc) represents the current waste management system in Kyoto for fiscal year 2008 (April 2008 to March 2009). In this scenario, the incineration facility recovers energy from waste in the form of electricity, but not heat. In the second scenario (SepBio), food and paper waste are separated at home, collected and sent to the facility, where they are digested anaerobically. We assume here that the wastes are treated by dry thermophilic anaerobic digestion. We also assume that the biogas from the anaerobic digestion is used to generate power and heat the methane fermenter, but not to supply heat outside the facility (e.g., district heating). The third scenario (MecBio) involves mechanical sorting of mixed household wastes at treatment facilities. Sorted wastes are then digested as in the SepBio scenario. In both the SepBio and the MecBio scenarios, the rest of the combustible waste is incinerated. We assumed that the amounts of the combustible waste to be treated in these three basic scenarios were the same (see Table 1).

In addition, we studied three waste reduction cases using SepBio as the baseline scenario. In the PrevLoss case, food waste is reduced by food loss prevention. In the ReducDrain case, waste reduction is achieved via water-draining methods such as the use of an outlet drain net or a sink corner strainer. In the ReducHcom case, food waste is reduced by home composting. In these three cases, we assumed that the amount of combustible waste would be reduced by 5%. In addition, we conducted a sensitivity analysis of the waste reduction rate to address the uncertainty associated with this parameter. The default assumption is based on our previous investigation of source-separated collection of food and paper waste in Kyoto from October 2008 to September 2009 (Matsuda et al. 2010b). In this previous study, we carried out questionnaire surveys regarding household disposal behavior and found that participants in the separate collection of food waste developed waste reduction behaviors such as food loss prevention and water draining. However, we did not have enough information to estimate the contribution of each waste reduction activity to the total 5% reduction in combustible waste. Thus, instead of analyzing one representative reduction case with mixed reduction activities, we compared three hypothetical waste reduction cases that each addressed one reduction activity.

The amounts of combustible waste for the three cases are shown in Table 1. To better represent the different reduction patterns of food waste among the three cases, we divided the food waste into four categories: cooking waste, leftovers, untouched food, and tea leaves and coffee residues. Here, cooking waste refers to the food waste generated during processing and preparation of meals such as fish bones and vegetable skins. In the PrevLoss case, only the food loss associated with leftovers and untouched food is reduced. In Kyoto, 37.4% of household combustible waste was food waste, of which 36.3% was food losses. Thus, food loss constitutes 13.6% (0.374×0.363) of the household combustible waste. Therefore, a 5% reduction in the combustible waste is equivalent to a 13.4% (0.05/0.374) reduction of food waste and 36.8% (0.05/0.136) reduction in food losses. We assumed that this reduction of food loss is achieved by reduced food production via constant food ingestion (i.e., not by constant food production with increased food ingestion). In the ReducDrain case, we assumed that the moisture content of the food waste was reduced from 77% to 73%, which is equivalent to a 13.4% reduction in food waste and a 5% reduction in combustible waste. In the ReducHcom case, we assumed that 13.4% of the food waste, or 5% of the combustible waste, is composted at home. The amounts of the food waste for the ReducDrain and ReducHcom cases were the same except for their moisture contents.

10.2.3 SYSTEM BOUNDARY

Figure 2 shows the system boundaries and the waste streams in this study. The food production process only depicts household food losses, i.e., food production processes for food ingested. All electricity generated from waste and biogas is transmitted through the power grids and replaces commercial electricity from utility companies, but is not used directly by the waste treatment facilities themselves. Compost substitutes for chemical fertilizer with the same amount of nitrogen. Construction, demolition, and final disposal of capital equipment were not considered in this study because they are relatively small (Matsuto et al. 2001). Thus, we decided to focus on the operating phase. In the ReducDrain case, wastewater from the kitchen would increase slightly; however, we did not consider the additional burden to waste water treatment because our preliminary assessment indicated that it made a small contribution to the total GHG emissions in the system being analyzed.

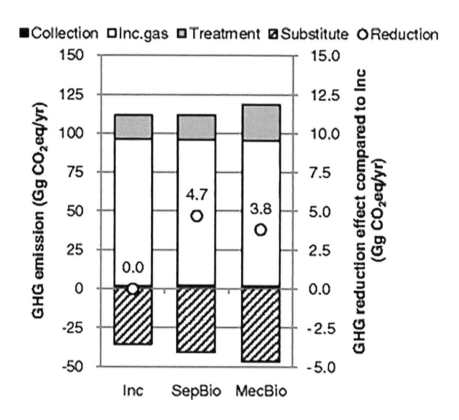

FIGURE 3: GHG emissions from the three waste treatment scenarios. The GHG emissions are divided into four processes: Collection, Inc. gas, Treatment, and Substitution. Inc. gas represents CO_2 emissions from the incineration of fossil fuel products in the waste. Reduction (white rectangles) indicate the GHG reduction compared to the Inc. gas scenario

10.2.4 IMPACT CATEGORY

Climate change was the only impact category considered for this study. In the 1990s, the major issues associated with waste management in Japan were landfill consumption and dioxin emissions; however, these have been intensively addressed over the past 20 years. In fact, the annual landfill consumption in Japan has decreased from 106 million tons/year in 1990 to 22 million tons/year in 2008 (Ministry of Environment 2011), and the dioxin emissions from waste incineration have decreased from 7200 g TEQ/year in 1997 to 100 g TEQ/year in 2009 (Ministry of Environment 2011). However, GHG emissions from the waste management sector in Japan have decreased only slightly, from 25.6 Tg CO_2-eq/year in 1990 to 21.8 Tg CO_2-eq/year in 2009 (GIO National Institute for Environmental Studies 2011). Thus, the authors believe that GHG emissions have become the most important environmental aspect for waste management in Japan. In addition, several environmental impact categories (e.g., acidification, smog formation) that are associated with emissions from thermal processes are generally strongly correlated with GHG emissions. This correlation suggests that GHG emissions could be used as a proxy indicator for the impact categories mentioned above.

It should be noted that the limited scope of the impact categories has some drawbacks. For example, nutrient emissions that lead to eutrophication were identified as an important impact category for food production processes (Davis and Sonesson 2008). However, the nutrient emissions are not strongly related to thermal processes, which invalidates the potential for use of GHG emissions as a proxy indicator. Despite these drawbacks, we focused on GHG emissions in this study. We will address the aforementioned drawbacks in future studies.

10.2.5 UNIT PROCESSES

In the food production process, we used calculated values (Matsuda et al. 2010a) to determine the GHG emissions from agricultural production, fishery activities, and processing and transportation of foods. Additionally, the emissions assigned to the leftovers included energy consumption for home cooking. It was assumed that all foods were grown and processed in Japan. This assumption is likely to underestimate the GHG emissions because the energy consumption for importing is not accounted for. Accordingly, the reduction in GHG emissions in the PrevLoss case might be underestimated.

In the collection process, we assumed that, in the separated collection scenarios (SepBio, PrevLoss, ReducDrain, ReducHcom), the separated waste and other combustible waste were each collected twice a week on different days. We also assumed that, in the nonseparated collection scenarios (Inc, MecBio), the combustible waste was collected twice a week. These assumptions are based on a separate collection experiment performed in Kyoto (Matsuda et al. 2010b). In the SepBio scenario, we applied the observed separation rates (Matsuda et al. 2010b) as the default values for these parameters. These data are listed in the Electronic Supplementary Material 1. We conducted a sensitivity analysis of the separation rate for the food waste to address the uncertainty associated with this parameter.

In the transport processes (expressed as arrows in Fig. 2), the average distance from the street side stations to the treatment facilities was 19 km, while that from the incineration facility to the landfill site was 50 km. These distances were based on actual geographical data for Kyoto.

In the mechanical sorting process, we applied published values for the sorting rates (Takuma 2005; see the Electronic Supplementary Material 1).

In the incineration process, electrical power consumption for the incinerator was calculated using an empirical formula (NIES 2008) based on waste composition. The CO_2 emissions associated with the combustion of each waste type were calculated using the elemental composition of the waste. Here, the carbon content of the biogenic waste produced by photosynthesis originates from the CO_2 in the atmosphere. Thus, CO_2 emissions from the biogenic waste were assumed to be carbon neutral and their net CO_2 emissions were considered to be zero. This accounting method was also applied to combustion of the biogas.

In the anaerobic digestion process, the biogas production was calculated based on the observed values in a pilot-scale study of Kyoto. Energy consumption in the anaerobic digestion facility was assumed to depend on the dry mass of the waste, because water is routinely added to the digestion reactor so that the water content of the waste remains constant. The digestate was incinerated in the incineration process.

In the landfill process, energy consumption for leachate treatment was calculated. We assumed that there were no landfill gas emissions, because only incineration residues with very low carbon content were landfilled in this study.

The home composting process was based on Tabata et al. (2009). We assumed that 10% of households used electric drying machines for composting, while the remaining 90% used nonelectric composters.

The global warming potentials of methane and nitrous oxide for this study were 25 and 298, respectively (IPCC 2007). Other parameters used in this study and some formulas are shown in the Electronic Supplementary Material 2 and 3 (Matsuda et al. 2010a; Murata 2000; NIES 2006; Tabata et al. 2009; Takuma 2005).

10.3 RESULTS

Figure 3 presents the GHG emissions from the three scenarios. Anaerobic digestion with separated collection (SepBio) was found to reduce GHG by 4.70 Gg CO_2-eq/year when compared with the incineration (Inc) scenario. Mechanical sorting and anaerobic digestion (MecBio) reduced GHG by 3.81 Gg CO_2-eq/year. The separation rate of the mechanical sorting is higher than that of the source separation in the SepBio scenario, enabling more power generation. However, MecBio consumes more energy during sorting of the mixed waste and treatment of nonbiodegradable waste than the separate collection does. Thus, the MecBio scenario was less favorable than the SepBio scenario with respect to GHG emissions.

Figure 4 shows the results for the waste reduction cases. The GHG emissions for food losses in Kyoto were estimated to be 46.8 Gg CO_2-eq/year. The GHG emissions were reduced by 21.5 Gg CO_2-eq/year in the PrevLoss case when compared to the Inc scenario. This reduction is more than four times larger than that of the SepBio scenario. In contrast, the GHG reduction in response to the ReducDrain case was almost the same as that for the SepBio scenario. This was because only the moisture content of the food waste was reduced, while the dry mass remained the same in the ReducDrain case. The energy consumption in the anaerobic digestion facility depends on the dry mass of the waste in our model; thus, the energy consumption is virtually unchanged. Finally, in the ReducHcom case, the GHG emissions increased by 2.60 Gg CO_2-eq/year when compared to the SepBio scenario. This was because the biogas production was reduced while the energy consumption was increased by the electric home composters. In addition, the reduction of GHG emissions by replacing chemical fertilizers with compost in the ReducHcom case was much less than that by the use of anaerobic digestion.

10.4 DISCUSSION

10.4.1 SENSITIVITY ANALYSIS

We conducted two sensitivity analyses to check the stability of the results and compare the relative importance of the separation rate and the waste reduction rate.

First, the sensitivity to the source separation rate was calculated (Fig. 5). To accomplish this, we fixed the separation rate for the paper waste at the default value and varied the separation rate for the food waste between 10% and 100%. The results showed that, even if all food waste was collected separately, the GHG reduction of the food losses prevention (PrevLoss) scenario was higher than that of the source-separated collection followed by biogasification without food waste reduction (SepBio).

Second, the sensitivity to the household waste reduction rate was calculated (Fig. 6). We varied the reduction rate of the household combustible waste from 0% to 10%, which corresponds to 0% to 27.0% of the food waste and 0% to 73.5% of the food loss. When the household waste was reduced by 10%, the GHG reduction of the PrevLoss, PrevDrain and PrevHcom cases was 34.2 Gg CO_2-eq/year, 1.0 Gg CO_2-eq/year, and -4.2 Gg CO_2-eq/year, respectively.

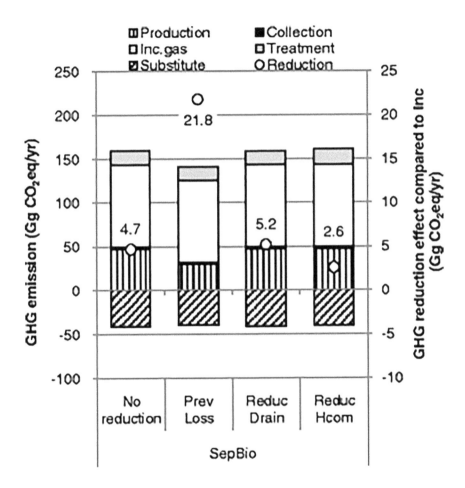

FIGURE 4: GHG emissions from the separated collection (SepBio) scenario and the three waste reduction cases. Production indicates the GHG emissions from production, distribution, and cooking for the food losses. Reduction (white rectangles) indicate the GHG reduction compared to the Inc. gas scenario

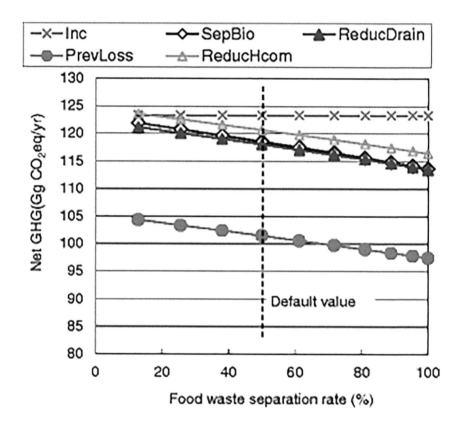

FIGURE 5: Sensitivity analysis of food waste source separation rate

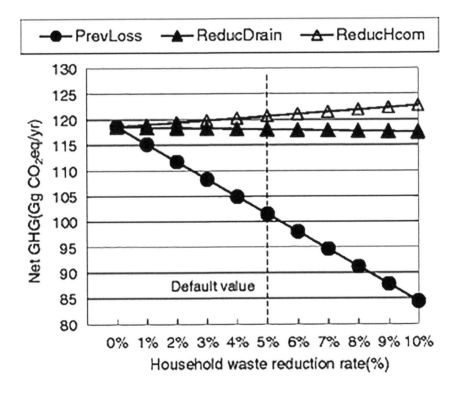

FIGURE 6: Sensitivity analysis of waste reduction rate

Next, we used these results to assess the impact of uncertainty regarding the relative contributions of the three waste reduction activities. To accomplish this, we fixed the sum of the waste reduction rates at 5%, changed the contributions from the three activities, and calculated the GHG reduction. The results are shown in a ternary contour graph (Fig. 7). The area within the triangle represents all possible combinations of the three activities that result in a total waste reduction of 5% (PrevLoss + ReducDrain + ReducHcom = 5%). The results show that GHG emissions will be reduced if the prevention of food loss accounts for more than one tenth of the total waste reduction by the three activities.

Comparison of Figs. 5 and 6 revealed that a reduction of only 1% of the household waste by food loss prevention has the same GHG reduction effect as a 31-point increase (from 50% to 81%) in the food waste separation rate. This comparison shows that a precise estimation of the food loss prevention rate is more important than that of the food waste separation rate to refine the estimated GHG reduction for the entire lifecycle of the source-separated collection systems.

10.4.2 COMPARISON WITH PREVIOUS STUDIES

We compared the results for the three waste management scenarios without waste reduction with those of previous studies. When the unit of the lifecycle GHG emissions was converted to per ton of waste, the results of the Inc, SepBio, and MecBio scenarios became 566 kg CO_2-eq/t waste, 545 kg CO_2-eq/t waste, and 549 kg CO_2-eq/t waste, respectively. Eriksson et al. (2005) demonstrated that the GHG emissions from the incineration of municipal solid waste were 855 kg CO_2-eq/t waste, while those from anaerobic digestion were 793 kg CO_2-eq/t waste. Inaba et al. (2010) reported that 331 kg CO_2-eq/t waste were emitted in the incineration scenario, while 325 kg CO_2-eq/t waste were emitted in the anaerobic digestion scenario. Our results are consistent with those of previous studies in that the GHG emissions from the anaerobic digestion scenarios are lower than those from the incineration scenarios. The differences in the absolute emission levels among these studies are likely a result of differences in the system boundaries, waste composition, and separation rates among studies.

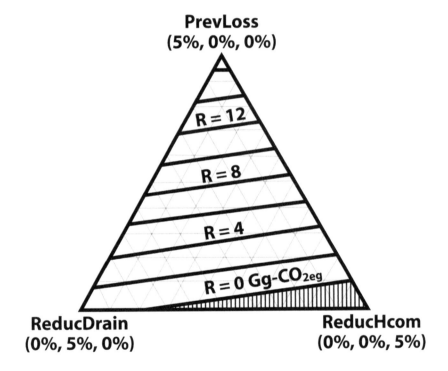

FIGURE 7: Reduction of GHG emissions in response to combinations of waste reduction activities: prevention of food loss (PrevLoss), draining of water (ReducDrain), and home composting (ReducHcom). The sum of the waste reduction rates in response to the three activities is fixed at 5%. Each of the three apexes of the ternary graph represents a situation in which only one of the three activities is practiced. The contour lines represent the level of GHG reductions. The shaded area shows the conditions under which the waste reduction will increase the GHG emissions

It is important to discuss the units of the GHG emissions. Specifically, the use of kg CO_2-eq/t should be avoided in favor of kg CO_2-eq/(person–year) in comparisons among several waste prevention scenarios from different studies, because the denominator of the former is affected by waste prevention while that of the latter is not. In waste prevention cases, the GHG emissions and amounts of waste are different from the reference scenario. In fact, a waste prevention case with lower GHG emissions than the reference scenario may result in higher GHG emissions per ton of waste than the default scenario (e.g., a case with 50% waste reduction and 40% GHG reduction will result in 120% (=60/50) of the reference GHG emissions per ton of waste).

10.4.3 LIMITATIONS

It should be noted that this study has the following limitations:

1. Only the GHG emissions are addressed.
2. There is great uncertainty associated with the waste reduction rate.
3. Contributions of the three reduction activities to the waste reduction are unknown.
4. Food loss prevention is assumed to be achieved by decreased production of food, not by increased ingestion of food.

Regarding the first limitation, previous studies have suggested the importance of nutrient emissions from food production processes. If we had included this impact category, the results would have reemphasized the importance of food loss prevention. Accordingly, the scope of the impact categories should be expanded to confirm the robustness of our results. Regarding the second limitation, our sensitivity analysis showed that even when the waste reduction rate is small, we can expect a large GHG reduction from the food loss prevention. Regarding the third limitation, our results showed that the three waste reduction activities have different effects on the GHG emissions, which suggests the importance of quantitative estimates to the contribution of each waste reduction activity. This point is one of the central targets of our ongoing research. The fourth limitation can be addressed in part by adding one more case, food loss prevention by increased ingestion. The results for this case would fall between those of the PrevLoss case and the ReducDrain case, suggesting that, from the viewpoint of global warming, food loss prevention is better achieved by decreased food production than increased ingestion.

10.4.4 IMPLICATIONS

Our results have two implications. First, it is necessary to advance the methods used for measuring changes in the disposal/consumption behavior of residents to quantify the impact of source-separated collection. Second, local authorities should expand the system boundaries for their strategic environmental assessment beyond waste treatment processes.

With respect to the first implication, precise estimations of the food waste separation rate and the waste reduction rate are important to better understand the impact of source-separated collection. In addition, contributions to the waste reduction from food loss prevention, draining, and home composting should be quantified. However, the waste reduction rate and the contribution of each activity are currently difficult to measure.

Measurement of the waste reduction rate is possible by comparing the amount of waste before and after introduction of source-separated collection. However, the effects of the source-separated collection should be distinguished from other factors such as annual fluctuations and long-term trends. Moreover, it is not easy to apply this method to elucidate the current reduction levels in municipalities in which source-separated collection has long been employed. To improve this estimation, accumulation of the empirical data and meta-analysis of these data would be useful. Panel data analysis might be especially useful for revealing the average waste reduction rate as well as the effects of different policy measures among local governments (e.g., collection frequency for the combustible waste) on the waste reduction rate. This type of empirical analysis has been widely applied to the pay-as-you-throw system (Miranda and Aldy 1998; Kinnaman and Fullerton 2000; Usui 2008). Accordingly, a similar approach to the separate collection system is warranted.

The measurement of the contribution of each waste reduction activity is more difficult. It is theoretically possible to determine the contributions of the three activities by developing a linear equations system that describes the changes in moisture content, food loss, and total food waste using three unknown parameters for the levels of the three waste reduction activities (see Electronic Supplementary Material 4). However, the estimates obtained by this method would likely be imprecise due to measurement errors during the waste composition analysis. To address the measurement error problem, estimation based on information sources other than the waste mass and waste composition would be useful. For example, surveys of the usage patterns of composters would enable quantification of the amount of home composting. A household expenditure survey might be a useful source for estimating changes in the amount of food purchased. Furthermore, this issue is not limited to food waste prevention. For example, reduction of plastic bottle waste will be achieved by weight saving, tumbler usage, and reduced consumption. Sharp et al. (2010) reviewed several methods for measuring the waste prevention effect and recommended combinations of these methods.

Regarding the second implication, our results confirmed the importance of the food production phase to the evaluation of food waste management systems. Thus, expansion of system boundaries beyond the waste treatment processes is inevitable. The US EPA (2010) developed the Waste Reduction Model to help solid waste planners calculate reductions in GHG emissions in response to several different waste management practices, including source reduction; however, the current version does not include emissions from food production and distribution due to data availability. The Ministry of the Environment of Japan published guidelines for strategic environmental assessment of waste management systems, but they do not cover those areas (MOE 2007a, b, 2008, 2011). Revision of these guidelines will likely encourage more local governments to attempt source-separated collection of food waste.

Finally, it should be noted that the perception of residents and public relations are important to the promotion of food loss prevention.

10.5 CONCLUSIONS

In this study, we focused on waste reduction activities and separate collection of household food and paper waste. The results showed that the effects of food waste reduction with separate collection on GHG emissions depend on the reason for the reduction in waste. We analyzed three cases that led to reduced waste for different reasons. In the first case, preventing food loss (PrevLoss) reduced significant amounts of GHG emissions. However, reducing the moisture contents of waste (ReducDrain) resulted in a much smaller GHG reduction. Finally, home composting (ReducHcom) increased GHG emissions. Therefore, to evaluate separate collection of household food waste, waste reduction activities should be considered. Food loss prevention has larger GHG reduction effects than anaerobic digestion. Therefore, the relationship between food loss prevention and waste management policies including separate collection will be the focus of our future research.

REFERENCES

5. Asano A, Yanase K, Takeda H, Mitsui M (2005) Study of the food waste recycling in Yokosuka City (third phase report). Proceedings of the 16th Annual Conference of the Japan Society of Waste Management Experts, Sendai (Japan), October 31–November 2, 2005, 1:484–486 (in Japanese)
6. Cleary J (2010) The incorporation of waste prevention activities into life cycle assessments of municipal solid waste management systems: methodological issues. Int J Life Cycle Assess 15(6):579–589
7. Cuellar AD, Webber ME (2010) Wasted food, wasted energy the embedded energy in food waste in the United States. Environ Sci Technol 44:6464–6469
8. Davis J, Sonesson U (2008) Life cycle assessment of integrated food chains - a Swedish case study of two chicken meals. Int J Life Cycle Assess 13:574–584
9. European Commission (2008) Green paper on the management of bio-waste in the European Union, COM (2008) 811 final, Brussels, 3.12.2008
10. European Commission (2010) Assessment of the options to improve the management of bio-waste in the European Union:1–237
11. Eriksson O, Reich MC, Frostell B, Björklund A, Assefa G, Sundqvist JO, Granath J, Baky A, Thyselius L (2005) Municipal solid waste management from a systems perspective. J Clean Prod 13:241–252

12. Fukushima K, Onoue Y, Nagao N, Niwa C, Toda T (2008) Life cycle assessment of anaerobic and aerobic biological treatment process for organic waste treatment. The 8th International Conference on EcoBalance, Tokyo (Japan), December 10–12, 2008,P-039

13. GIO National Institute for Environmental Studies (NIES), Japan (2011) The GHGs emissions data of Japan (1990–2009). http://www-gio.nies.go.jp/aboutghg/nir/nir-e. html. Accessed September 18, 2011

14. Hirose Y (1995) Social psychology for environment and consumption. Nagoya University Press, Nagoya

15. Inaba R, Nansai K, Fujii M, Hashimoto S (2010) Hybrid life-cycle assessment (LCA) of CO2 emission alternatives for household food wastes in Japan. Waste Manage Res 28(6):496–507

16. IPCC (2007) Climate change 2007 (AR4), Working group I report. The physical science basis

17. Kim MH, Kim JW (2010) Comparison through a LCA evaluation analysis of food waste disposal options from the perspective of global warming and resource recovery. Sci Total Environ 408(19):3998–4006

18. Kinnaman TC, Fullerton D (2000) Garbage and recycling with endogenous local policy. J Urban Econ 48(3):419–442

19. Kyoto City Environmental Policy Bureau (2009a) Survey of detailed household waste composition in Kyoto FY2008 (in Japanese)

20. Kyoto City Environmental Policy Bureau (2009b) Operation summary of environmental policy bureau FY2009 (in Japanese). http://www.city.kyoto.lg.jp/kankyo/page/0000072597.html. Accessed December 11, 2010

21. Matsuda T, Yano J, Hirai Y, Sakai S, Yamada K, Ogiuchi M, Hori H (2010a) Life cycle analysis of household waste management considering trade-off between food waste reduction and recycling. J Life Cycle Ass Japan 6(4):280–287 (in Japanese)

22. Matsuda T, Yano J, Hirai Y, Yamada K, Ogiuchi M, Hori H, Sakai S (2010b) Life cycle inventory analysis on biogasification of household food and paper waste. 9th International conference on ecobalance, Tokyo (Japan), December 9–12, 2010, pp D1–1610

23. Matsuto T, Tanaka N, Habara H (2001) Computer modeling to assess cost, energy consumption and carbon dioxide emission in zero emission scenario of municipal solid waste. 12th Jpn Soc Waste Manage Exp, pp 134–146

24. Ministry of the Environment (MOE), Japan (2007a) Guidelines for introducing strategic environmental assessment

25. Ministry of the Environment (MOE), Japan (2007b) Guidelines on municipal solid waste management systems toward a sound material-cycle society for local governments

26. Ministry of the Environment (MOE), Japan (2008) Survey of the municipal waste treatment FY2008 (in Japanese). http://www.env.go.jp/recycle/waste_tech/ippan/h20/index.html. Accessed December 11, 2010

27. Ministry of the Environment (MOE), Japan (2011) Environmental statistics 2011 (in Japanese). http://www.env.go.jp/doc/toukei/index.html. Accessed September 18, 2011

28. Miranda ML, Aldy JE (1998) Unit pricing of residential municipal solid waste: lessons from nine case study communities. J Environ Manage 52(1):79–93
29. Murata M (2000) Life cycle assessment for food waste recycling and management (in Japanese). Kyoto University master's thesis
30. NIES, Japan (2006) Technical development report of hydrogen production from biomass and biowaste in FY2005 (in Japanese). NIES, Tsukuba, pp 7.193
31. NIES, Japan (2008) Technical development report of hydrogen production from biomass and biowaste in FY2007 (in Japanese). NIES, Tsukuba, pp 7.44–7.63
32. Sakai S, Hirai Y, Yoshikawa K, Deguchi S (2005) Distribution of potential biomass/waste resource and GHG emission analysis for food waste recycling systems (in Japanese). J Japan Soc Waste Manage Experts 16(2):173–187
33. Sharp V, Giorgi S, Wilson DC (2010) Methods to monitor and evaluate household waste prevention. Waste Manage Res 28(3):269–280
34. Sonesson U, Björklund A, Carlsson M, Dalemo M (2000) Environmental and economic analysis of management systems for biodegradable waste. Resour Conserv Recy 28:29–53
35. Tabata T, Ihara T, Nakazawa H, Genchi Y (2009) Evaluating the self-disposal of household waste in regional towns and cities: Present situation and analysis of environmental and economic effects. J Jpn Soc Mater Cycles Waste Manage 20(2):99–110 (in Japanese)
36. Takuma (2005) The development of efficient energy and material recovery technology by dry methane fermentation. Grants for development of future waste treatment technology (in Japanese). Technical report, pp 1–29
37. Tatara M, Kitajima Y, Kikuchi S, Yoshimura Y, Maeda H, Okabe M, Miyano H, Kuramochi K, Murayama H, Endoh T (2010) Development of a biogasification system for unsorted domestic combustible waste (in Japanese). Kajima Tech Res Inst Annu Rep 58:121–126
38. US EPA (2010) Waste reduction model (WARM) http://www.epa.gov/climatechange/wycd/waste/calculators/Warm_home.html. Accessed September 27, 2011
39. Usui T (2008) Estimating the effect of unit-based pricing in the presence of sample selection bias under Japanese recycling law. Ecol Econ 66(2–3):282–288
40. WRAP (2009) Household Food and Drink Waste in the UK, http://www.wrap.org.uk/retail_supply_chain/research_tools/research/report_household.html. Accessed December 11, 2010

There are several supplemental files that are not available in this version of the article. To view this additional information, please use the citation on the first page of this chapter.

CHAPTER 11

Assessing "Dangerous Climate Change": Required Reduction of Carbon Emissions to Protect Young People, Future Generations and Nature

JAMES HANSEN, PUSHKER KHARECHA, MAKIKO SATO, VALERIE MASSON-DELMOTTE, FRANK ACKERMAN, DAVID J. BEERLING, PAUL J. HEARTY, OVE HOEGH-GULDBERG, SHI-LING HSU, CAMILLE PARMESAN, JOHAN ROCKSTROM, EELCO J. ROHLING, JEFFREY SACHS, PETE SMITH, KONRAD STEFFEN, LISE VAN SUSTEREN, KARINA VON SCHUCKMANN, AND JAMES C. ZACHOS

11.1 INTRODUCTION

Humans are now the main cause of changes of Earth's atmospheric composition and thus the drive for future climate change [1]. The principal climate forcing, defined as an imposed change of planetary energy balance [1]–[2], is increasing carbon dioxide (CO_2) from fossil fuel emissions, much of which will remain in the atmosphere for millennia [1], [3]. The

Assessing "Dangerous Climate Change": Required Reduction of Carbon Emissions to Protect Young People, Future Generations and Nature. © Hansen J et al. PLoS ONE 8,12 (2013). http://journals. plos.org/plosone/article?id=10.1371/journal.pone.0081648. The work is made available under the Creative Commons CC0 public domain dedication, http://creativecommons.org/publicdomain/ zero/1.0/.

climate response to this forcing and society's response to climate change are complicated by the system's inertia, mainly due to the ocean and the ice sheets on Greenland and Antarctica together with the long residence time of fossil fuel carbon in the climate system. The inertia causes climate to appear to respond slowly to this human-made forcing, but further long-lasting responses can be locked in.

More than 170 nations have agreed on the need to limit fossil fuel emissions to avoid dangerous human-made climate change, as formalized in the 1992 Framework Convention on Climate Change [6]. However, the stark reality is that global emissions have accelerated (Fig. 1) and new efforts are underway to massively expand fossil fuel extraction [7]–[9] by drilling to increasing ocean depths and into the Arctic, squeezing oil from tar sands and tar shale, hydro-fracking to expand extraction of natural gas, developing exploitation of methane hydrates, and mining of coal via mountaintop removal and mechanized long-wall mining. The growth rate of fossil fuel emissions increased from 1.5%/year during 1980–2000 to 3%/year in 2000–2012, mainly because of increased coal use [4]–[5].

The Framework Convention [6] does not define a dangerous level for global warming or an emissions limit for fossil fuels. The European Union in 1996 proposed to limit global warming to 2°C relative to pre-industrial times [10], based partly on evidence that many ecosystems are at risk with larger climate change. The 2°C target was reaffirmed in the 2009 "Copenhagen Accord" emerging from the 15th Conference of the Parties of the Framework Convention [11], with specific language "We agree that deep cuts in global emissions are required according to science, as documented in the IPCC Fourth Assessment Report with a view to reduce global emissions so as to hold the increase in global temperature below 2 degrees Celsius…".

A global warming target is converted to a fossil fuel emissions target with the help of global climate-carbon-cycle models, which reveal that eventual warming depends on cumulative carbon emissions, not on the temporal history of emissions [12]. The emission limit depends on climate sensitivity, but central estimates [12]–[13], including those in the upcoming Fifth Assessment of the Intergovernmental Panel on Climate Change [14], are that a 2°C global warming limit implies a cumulative carbon emissions limit of the order of 1000 GtC. In comparing carbon emissions, note that some authors emphasize the sum of fossil fuel and deforestation carbon. We bookkeep fossil fuel and deforestation carbon separately, because the larger fossil fuel term is known more accurately and this carbon stays in the climate system for hundreds of thousands of years. Thus fossil fuel carbon is the crucial human input that must be limited. Deforestation carbon is more uncertain and potentially can be offset on the century time scale by storage in the biosphere, including the soil, via reforestation and improved agricultural and forestry practices.

There are sufficient fossil fuel resources to readily supply 1000 GtC, as fossil fuel emissions to date (370 GtC) are only a small fraction of potential emissions from known reserves and potentially recoverable resources (Fig. 2). Although there are uncertainties in reserves and resources, ongoing fossil fuel subsidies and continuing technological advances ensure that more and more of these fuels will be economically recoverable. As we will show, Earth's paleoclimate record makes it clear that the CO_2 produced by burning all or most of these fossil fuels would lead to a very different planet than the one that humanity knows.

Our evaluation of a fossil fuel emissions limit is not based on climate models but rather on observational evidence of global climate change as a function of global temperature and on the fact that climate stabilization requires long-term planetary energy balance. We use measured global temperature and Earth's measured energy imbalance to determine the atmospheric CO_2 level required to stabilize climate at today's global temperature, which is near the upper end of the global temperature range in the current interglacial period (the Holocene). We then examine climate impacts during the past few decades of global warming and in paleoclimate records including the Eemian period, concluding that there are already clear indications of undesirable impacts at the current level of warming and that 2°C warming would have major deleterious consequences. We use simple representations of the carbon cycle and global temperature, consistent with observations, to simulate transient global temperature and assess carbon emission scenarios that could keep global climate near the Holocene range. Finally, we discuss likely overshooting of target emissions, the potential for carbon extraction from the atmosphere, and implications for energy and economic policies, as well as intergenerational justice.

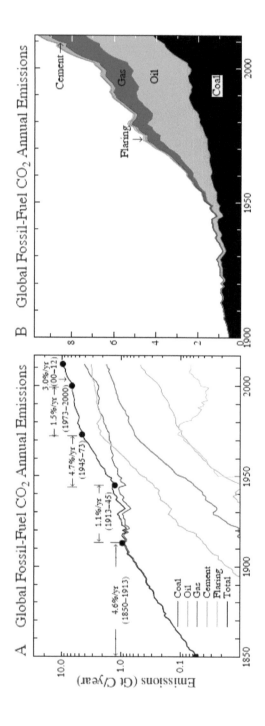

FIGURE 1: CO_2 annual emissions from fossil fuel use and cement manufacture, based on data of British Petroleum [4] concatenated with data of Boden et al. [5]. (A) is log scale and (B) is linear.

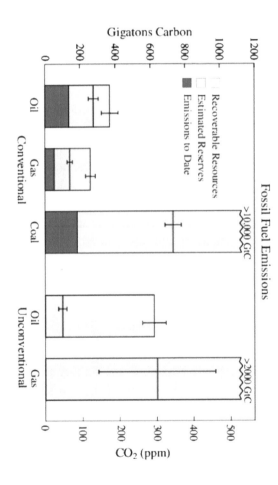

FIGURE 2: Fossil fuel CO_2 emissions and carbon content (1 ppm atmospheric CO_2 ~ 2.12 GtC). Estimates of reserves (profitable to extract at current prices) and resources (potentially recoverable with advanced technology and/or at higher prices) are the mean of estimates of Energy Information Administration (EIA) [7], German Advisory Council (GAC) [8], and Global Energy Assessment (GEA) [9]. GEA [9] suggests the possibility of >15,000 GtC unconventional gas. Error estimates (vertical lines) are from GEA and probably underestimate the total uncertainty. We convert energy content to carbon content using emission factors of Table 4.2 of [15] for coal, gas and conventional oil, and, also following [15], emission factor of unconventional oil is approximated as being the same as for coal. Total emissions through 2012, including gas flaring and cement manufacture, are 384 GtC; fossil fuel emissions alone are ~370 GtC.

11.2 GLOBAL TEMPERATURE AND EARTH'S ENERGY BALANCE

Global temperature and Earth's energy imbalance provide our most useful measuring sticks for quantifying global climate change and the changes of global climate forcings that would be required to stabilize global climate. Thus we must first quantify knowledge of these quantities.

11.2.1 TEMPERATURE

Temperature change in the past century (Fig. 3; update of figures in [16]) includes unforced variability and forced climate change. The long-term global warming trend is predominantly a forced climate change caused by increased human-made atmospheric gases, mainly CO_2 [1]. Increase of "greenhouse" gases such as CO_2 has little effect on incoming sunlight but makes the atmosphere more opaque at infrared wavelengths, causing infrared (heat) radiation to space to emerge from higher, colder levels, which thus reduces infrared radiation to space. The resulting planetary energy imbalance, absorbed solar energy exceeding heat emitted to space, causes Earth to warm. Observations, discussed below, confirm that Earth is now substantially out of energy balance, so the long-term warming will continue.

Global temperature appears to have leveled off since 1998 (Fig. 3a). That plateau is partly an illusion due to the 1998 global temperature spike caused by the El Niño of the century that year. The 11-year (132-month) running mean temperature (Fig. 3b) shows only a moderate decline of the warming rate. The 11-year averaging period minimizes the effect of variability due to the 10–12 year periodicity of solar irradiance as well as irregular El Niño/La Niña warming/cooling in the tropical Pacific Ocean. The current solar cycle has weaker irradiance than the several prior solar cycles, but the decreased irradiance can only partially account for the decreased warming rate [17]. Variability of the El Niño/La Niña cycle, described as a Pacific Decadal Oscillation, largely accounts for the temporary decrease of warming [18], as we discuss further below in conjunction with global temperature simulations.

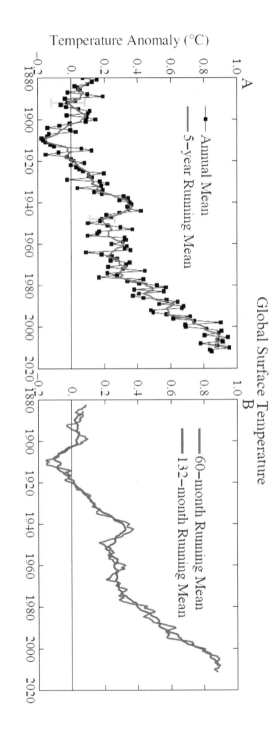

FIGURE 3: Global surface temperature relative to 1880–1920 mean. B shows the 5 and 11 year means. Figures are updates of [16] using data through August 2013.

Assessments of dangerous climate change have focused on estimating a permissible level of global warming. The Intergovernmental Panel on Climate Change [1], [19] summarized broad-based assessments with a "burning embers" diagram, which indicated that major problems begin with global warming of 2–3°C. A probabilistic analysis [20], still partly subjective, found a median "dangerous" threshold of 2.8°C, with 95% confidence that the dangerous threshold was 1.5°C or higher. These assessments were relative to global temperature in year 1990, so add 0.6°C to these values to obtain the warming relative to 1880–1920, which is the base period we use in this paper for preindustrial time. The conclusion that humanity could tolerate global warming up to a few degrees Celsius meshed with common sense. After all, people readily tolerate much larger regional and seasonal climate variations.

The fallacy of this logic emerged recently as numerous impacts of ongoing global warming emerged and as paleoclimate implications for climate sensitivity became apparent. Arctic sea ice end-of-summer minimum area, although variable from year to year, has plummeted by more than a third in the past few decades, at a faster rate than in most models [21], with the sea ice thickness declining a factor of four faster than simulated in IPCC climate models [22]. The Greenland and Antarctic ice sheets began to shed ice at a rate, now several hundred cubic kilometers per year, which is continuing to accelerate [23]–[25]. Mountain glaciers are receding rapidly all around the world [26]–[29] with effects on seasonal freshwater availability of major rivers [30]–[32]. The hot dry subtropical climate belts have expanded as the troposphere has warmed and the stratosphere cooled [33]–[36], contributing to increases in the area and intensity of drought [37] and wildfires [38]. The abundance of reef-building corals is decreasing at a rate of 0.5–2%/year, at least in part due to ocean warming and possibly ocean acidification caused by rising dissolved CO_2 [39]–[41]. More than half of all wild species have shown significant changes in where they live and in the timing of major life events [42]–[44]. Mega-heatwaves, such as those in Europe in 2003, the Moscow area in 2010, Texas and Oklahoma in 2011, Greenland in 2012, and Australia in 2013 have become more widespread with the increase demonstrably linked to global warming [45]–[47].

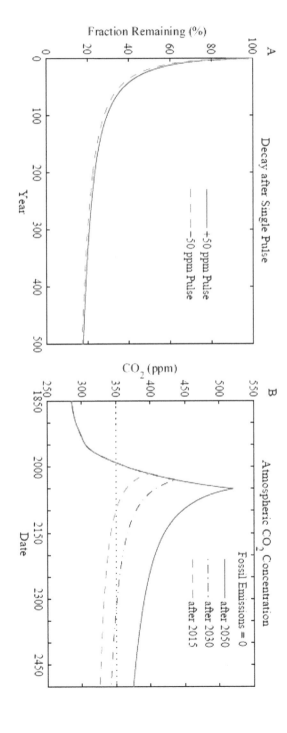

FIGURE 4: Decay of atmospheric CO_2 perturbations. (A) Instantaneous injection or extraction of CO_2 with initial conditions at equilibrium. (B) Fossil fuel emissions terminate at the end of 2015, 2030, or 2050 and land use emissions terminate after 2015 in all three cases, i.e., thereafter there is no net deforestation.

These growing climate impacts, many more rapid than anticipated and occurring while global warming is less than 1°C, imply that society should reassess what constitutes a "dangerous level" of global warming. Earth's paleoclimate history provides a valuable tool for that purpose.

11.2.2 PALEOCLIMATE TEMPERATURE

Major progress in quantitative understanding of climate change has occurred recently by use of the combination of data from high resolution ice cores covering time scales of order several hundred thousand years [48]–[49] and ocean cores for time scales of order one hundred million years [50]. Quantitative insights on global temperature sensitivity to external forcings [51]–[52] and sea level sensitivity to global temperature [52]–[53] are crucial to our analyses. Paleoclimate data also provide quantitative information about how nominally slow feedback processes amplify climate sensitivity [51]–[52], [54]–[56], which also is important to our analyses.

Earth's surface temperature prior to instrumental measurements is estimated via proxy data. We will refer to the surface temperature record in Fig. 4 of a recent paper [52]. Global mean temperature during the Eemian interglacial period (120,000 years ago) is constrained to be 2°C warmer than our pre-industrial (1880–1920) level based on several studies of Eemian climate [52]. The concatenation of modern and instrumental records [52] is based on an estimate that global temperature in the first decade of the 21st century (+0.8°C relative to 1880–1920) exceeded the Holocene mean by 0.25±0.25°C. That estimate was based in part on the fact that sea level is now rising 3.2 mm/yr (3.2 m/millennium) [57], an order of magnitude faster than the rate during the prior several thousand years, with rapid change of ice sheet mass balance over the past few decades [23] and Greenland and Antarctica now losing mass at accelerating rates [23]–[24]. This concatenation, which has global temperature 13.9°C in the base period 1951–1980, has the first decade of the 21st century slightly (~0.1°C) warmer than the early Holocene maximum. A recent reconstruction from proxy temperature data [55] concluded that global temperature declined about 0.7°C between the Holocene maximum and a pre-industrial minimum before recent warming brought temperature back near the Holocene maximum, which is consistent with our analysis.

Climate oscillations evident in Fig. 4 of Hansen et al. [52] were instigated by perturbations of Earth's orbit and spin axis tilt relative to the orbital plane, which alter the geographical and seasonal distribution of sunlight on Earth [58]. These forcings change slowly, with periods between 20,000 and 400,000 years, and thus climate is able to stay in quasi-equilibrium with these forcings. Slow insolation changes initiated the climate oscillations, but the mechanisms that caused the climate changes to be so large were two powerful amplifying feedbacks: the planet's surface albedo (its reflectivity, literally its whiteness) and atmospheric CO_2 amount. As the planet warms, ice and snow melt, causing the surface to be darker, absorb more sunlight and warm further. As the ocean and soil become warmer they release CO_2 and other greenhouse gases, causing further warming. Together with fast feedbacks processes, via changes of water vapor, clouds, and the vertical temperature profile, these slow amplifying feedbacks were responsible for almost the entire glacial-to-interglacial temperature change [59]–[62].

The albedo and CO_2 feedbacks amplified weak orbital forcings, the feedbacks necessarily changing slowly over millennia, at the pace of orbital changes. Today, however, CO_2 is under the control of humans as fossil fuel emissions overwhelm natural changes. Atmospheric CO_2 has increased rapidly to a level not seen for at least 3 million years [56], [63]. Global warming induced by increasing CO_2 will cause ice to melt and hence sea level to rise as the global volume of ice moves toward the quasi-equilibrium amount that exists for a given global temperature [53]. As ice melts and ice area decreases, the albedo feedback will amplify global warming.

Earth, because of the climate system's inertia, has not yet fully responded to human-made changes of atmospheric composition. The ocean's thermal inertia, which delays some global warming for decades and even centuries, is accounted for in global climate models and its effect is confirmed via measurements of Earth's energy balance (see next section). In addition there are slow climate feedbacks, such as changes of ice sheet size, that occur mainly over centuries and millennia. Slow feedbacks have little effect on the immediate planetary energy balance, instead coming into play in response to temperature change. The slow feedbacks are difficult to model, but paleoclimate data and observations of ongoing changes help provide quantification.

11.2.3 EARTH'S ENERGY IMBALANCE

At a time of climate stability, Earth radiates as much energy to space as it absorbs from sunlight. Today Earth is out of balance because increasing atmospheric gases such as CO_2 reduce Earth's heat radiation to space, thus causing an energy imbalance, as there is less energy going out than coming in. This imbalance causes Earth to warm and move back toward energy balance. The warming and restoration of energy balance take time, however, because of Earth's thermal inertia, which is due mainly to the global ocean.

Earth warmed about 0.8°C in the past century. That warming increased Earth's radiation to space, thus reducing Earth's energy imbalance. The remaining energy imbalance helps us assess how much additional warming is still "in the pipeline". Of course increasing CO_2 is only one of the factors affecting Earth's energy balance, even though it is the largest climate forcing. Other forcings include changes of aerosols, solar irradiance, and Earth's surface albedo.

Determination of the state of Earth's climate therefore requires measuring the energy imbalance. This is a challenge, because the imbalance is expected to be only about 1 W/m² or less, so accuracy approaching 0.1 W/m² is needed. The most promising approach is to measure the rate of changing heat content of the ocean, atmosphere, land, and ice [64]. Measurement of ocean heat content is the most critical observation, as nearly 90 percent of the energy surplus is stored in the ocean [64]–[65].

11.2.4 OBSERVED ENERGY IMBALANCE

Nations of the world have launched a cooperative program to measure changing ocean heat content, distributing more than 3000 Argo floats around the world ocean, with each float repeatedly diving to a depth of 2 km and back [66]. Ocean coverage by floats reached 90% by 2005 [66], with the gaps mainly in sea ice regions, yielding the potential for an accurate energy balance assessment, provided that several systematic measurement biases exposed in the past decade are minimized [67]–[69].

Argo data reveal that in 2005–2010 the ocean's upper 2000 m gained heat at a rate equal to 0.41 W/m² averaged over Earth's surface [70]. Smaller contributions to planetary energy imbalance are from heat gain by the deeper ocean (+0.10 W/m²), energy used in net melting of ice (+0.05 W/m²), and energy taken up by warming continents (+0.02 W/m²). Data sources for these estimates and uncertainties are provided elsewhere [64]. The resulting net planetary energy imbalance for the six years 2005–2010 is +0.58±0.15 W/m².

The positive energy imbalance in 2005–2010 confirms that the effect of solar variability on climate is much less than the effect of human-made greenhouse gases. If the sun were the dominant forcing, the planet would have a negative energy balance in 2005–2010, when solar irradiance was at its lowest level in the period of accurate data, i.e., since the 1970s [64], [71]. Even though much of the greenhouse gas forcing has been expended in causing observed 0.8°C global warming, the residual positive forcing overwhelms the negative solar forcing. The full amplitude of solar cycle forcing is about 0.25 W/m² [64], [71], but the reduction of solar forcing due to the present weak solar cycle is about half that magnitude as we illustrate below, so the energy imbalance measured during solar minimum (0.58 W/m²) suggests an average imbalance over the solar cycle of about 0.7 W/m².

Earth's measured energy imbalance has been used to infer the climate forcing by aerosols, with two independent analyses yielding a forcing in the past decade of about −1.5 W/m² [64], [72], including the direct aerosol forcing and indirect effects via induced cloud changes. Given this large (negative) aerosol forcing, precise monitoring of changing aerosols is needed [73]. Public reaction to increasingly bad air quality in developing regions [74] may lead to future aerosol reductions, at least on a regional basis. Increase of Earth's energy imbalance from reduction of particulate air pollution, which is needed for the sake of human health, can be minimized via an emphasis on reducing absorbing black soot [75], but the potential to constrain the net increase of climate forcing by focusing on black soot is limited [76].

11.2.5 ENERGY IMBALANCE IMPLICATIONS FOR CO_2 TARGET

Earth's energy imbalance is the most vital number characterizing the state of Earth's climate. It informs us about the global temperature change "in the pipeline" without further change of climate forcings and it defines how much greenhouse gases must be reduced to restore Earth's energy balance, which, at least to a good approximation, must be the requirement for stabilizing global climate. The measured energy imbalance accounts for all natural and human-made climate forcings, including changes of atmospheric aerosols and Earth's surface albedo.

If Earth's mean energy imbalance today is +0.5 W/m^2, CO_2 must be reduced from the current level of 395 ppm (global-mean annual-mean in mid-2013) to about 360 ppm to increase Earth's heat radiation to space by 0.5 W/m^2 and restore energy balance. If Earth's energy imbalance is 0.75 W/m^2, CO_2 must be reduced to about 345 ppm to restore energy balance [64], [75].

The measured energy imbalance indicates that an initial CO_2 target "<350 ppm" would be appropriate, if the aim is to stabilize climate without further global warming. That target is consistent with an earlier analysis [54]. Additional support for that target is provided by our analyses of ongoing climate change and paleoclimate, in later parts of our paper. Specification now of a CO_2 target more precise than <350 ppm is difficult and unnecessary, because of uncertain future changes of forcings including other gases, aerosols and surface albedo. More precise assessments will become available during the time that it takes to turn around CO_2 growth and approach the initial 350 ppm target.

Below we find the decreasing emissions scenario that would achieve the 350 ppm target within the present century. Specifically, we want to know the annual percentage rate at which emissions must be reduced to reach this target, and the dependence of this rate upon the date at which reductions are initiated. This approach is complementary to the approach of estimating cumulative emissions allowed to achieve a given limit on global warming [12].

If the only human-made climate forcing were changes of atmospheric CO_2, the appropriate CO_2 target might be close to the pre-industrial CO_2 amount [53]. However, there are other human forcings, including aerosols, the effect of aerosols on clouds, non-CO_2 greenhouse gases, and changes of surface albedo that will not disappear even if fossil fuel burning is phased out. Aerosol forcings are substantially a result of fossil fuel burning [1], [76], but the net aerosol forcing is a sensitive function of various aerosol sources [76]. The indirect aerosol effect on clouds is non-linear [1], [76] such that it has been suggested that even the modest aerosol amounts added by pre-industrial humans to an otherwise pristine atmosphere may have caused a significant climate forcing [59]. Thus continued precise monitoring of Earth's radiation imbalance is probably the best way to assess and adjust the appropriate CO_2 target.

Ironically, future reductions of particulate air pollution may exacerbate global warming by reducing the cooling effect of reflective aerosols. However, a concerted effort to reduce non-CO_2 forcings by methane, tropospheric ozone, other trace gases, and black soot might counteract the warming from a decline in reflective aerosols [54], [75]. Our calculations below of future global temperature assume such compensation, as a first approximation. To the extent that goal is not achieved, adjustments must be made in the CO_2 target or future warming may exceed calculated values.

11.3 CLIMATE IMPACTS

Determination of the dangerous level of global warming inherently is partly subjective, but we must be as quantitative as possible. Early estimates for dangerous global warming based on the "burning embers" approach [1], [19]–[20] have been recognized as probably being too conservative [77]. A target of limiting warming to 2°C has been widely adopted, as discussed above. We suspect, however, that this may be a case of inching toward a better answer. If our suspicion is correct, then that gradual approach is itself very dangerous, because of the climate system's inertia. It will become exceedingly difficult to keep warming below a target smaller than 2°C, if high emissions continue much longer.

We consider several important climate impacts and use evidence from current observations to assess the effect of 0.8°C warming and paleoclimate data for the effect of larger warming, especially the Eemian period, which had global mean temperature about +2°C relative to pre-industrial time. Impacts of special interest are sea level rise and species extermination, because they are practically irreversible, and others important to humankind.

11.3.1 SEA LEVEL

The prior interglacial period, the Eemian, was at most ~2°C warmer than 1880–1920 (Fig. 3). Sea level reached heights several meters above today's level [78]–[80], probably with instances of sea level change of the order of 1 m/century [81]–[83]. Geologic shoreline evidence has been interpreted as indicating a rapid sea level rise of a few meters late in the Eemian to a peak about 9 meters above present, suggesting the possibility that a critical stability threshold was crossed that caused polar ice sheet collapse [84]–[85], although there remains debate within the research community about this specific history and interpretation. The large Eemian sea level excursions imply that substantial ice sheet melting occurred when the world was little warmer than today.

During the early Pliocene, which was only ~3°C warmer than the Holocene, sea level attained heights as much as 15–25 meters higher than today [53], [86]–[89]. Such sea level rise suggests that parts of East Antarctica must be vulnerable to eventual melting with global temperature increase of a few degrees Celsius. Indeed, satellite gravity data and radar altimetry reveal that the Totten Glacier of East Antarctica, which fronts a large ice mass grounded below sea level, is now losing mass [90].

Greenland ice core data suggest that the Greenland ice sheet response to Eemian warmth was limited [91], but the fifth IPCC assessment [14] concludes that Greenland very likely contributed between 1.4 and 4.3 m to the higher sea level of the Eemian. The West Antarctic ice sheet is probably more susceptible to rapid change, because much of it rests on bedrock well below sea level [92]–[93]. Thus the entire 3–4 meters of global sea level contained in that ice sheet may be vulnerable to rapid disintegration, although arguments for stability of even this marine ice sheet have been made [94]. However, Earth's history reveals sea level changes of as much as a few meters per century, even though the natural climate forcings changed much more slowly than the present human-made forcing.

Expected human-caused sea level rise is controversial in part because predictions focus on sea level at a specific time, 2100. Sea level on a given date is inherently difficult to predict, as it depends on how rapidly non-linear ice sheet disintegration begins. Focus on a single date also encourages people to take the estimated result as an indication of what humanity faces, thus failing to emphasize that the likely rate of sea level rise immediately after 2100 will be much larger than within the 21st century, especially if CO_2 emissions continue to increase.

Recent estimates of sea level rise by 2100 have been of the order of 1 m [95]–[96], which is higher than earlier assessments [26], but these estimates still in part assume linear relations between warming and sea level rise. It has been argued [97]–[98] that continued business-as-usual CO_2 emissions are likely to spur a nonlinear response with multi-meter sea level rise this century. Greenland and Antarctica have been losing mass at rapidly increasing rates during the period of accurate satellite data [23]; the data are suggestive of exponential increase, but the records are too short to be conclusive. The area on Greenland with summer melt has increased markedly, with 97% of Greenland experiencing melt in 2012 [99].

The important point is that the uncertainty is not about whether continued rapid CO_2 emissions would cause large sea level rise, submerging global coastlines—it is about how soon the large changes would begin. The carbon from fossil fuel burning will remain in and affect the climate system for many millennia, ensuring that over time sea level rise of many meters will occur—tens of meters if most of the fossil fuels are burned [53]. That order of sea level rise would result in the loss of hundreds of historical coastal cities worldwide with incalculable economic consequences, create hundreds of millions of global warming refugees from highly-populated low-lying areas, and thus likely cause major international conflicts.

11.3.2 SHIFTING CLIMATE ZONES

Theory and climate models indicate that the tropical overturning (Hadley) atmospheric circulation expands poleward with global warming [33].

There is evidence in satellite and radiosonde data and in observational data for poleward expansion of the tropical circulation by as much as a few degrees of latitude since the 1970s [34]–[35], but natural variability may have contributed to that expansion [36]. Change in the overturning circulation likely contributes to expansion of subtropical conditions and increased aridity in the southern United States [30], [100], the Mediterranean region, South America, southern Africa, Madagascar, and southern Australia. Increased aridity and temperature contribute to increased forest fires that burn hotter and are more destructive [38].

Despite large year-to-year variability of temperature, decadal averages reveal isotherms (lines of a given average temperature) moving poleward at a typical rate of the order of 100 km/decade in the past three decades [101], although the range shifts for specific species follow more complex patterns [102]. This rapid shifting of climate zones far exceeds natural rates of change. Movement has been in the same direction (poleward, and upward in elevation) since about 1975. Wild species have responded to climate change, with three-quarters of marine species shifting their ranges poleward as much as 1000 km [44], [103] and more than half of terrestrial species shifting ranges poleward as much as 600 km and upward as much as 400 m [104].

Humans may adapt to shifting climate zones better than many species. However, political borders can interfere with human migration, and indigenous ways of life already have been adversely affected [26]. Impacts are apparent in the Arctic, with melting tundra, reduced sea ice, and increased shoreline erosion. Effects of shifting climate zones also may be important for indigenous Americans who possess specific designated land areas, as well as other cultures with long-standing traditions in South America, Africa, Asia and Australia.

11.3.3 HUMAN EXTERMINATION OF SPECIES

Biodiversity is affected by many agents including overharvesting, introduction of exotic species, land use changes, nitrogen fertilization, and direct effects of increased atmospheric CO_2 on plant ecophysiology [43]. However, an overriding role of climate change is exposed by diverse effects of rapid warming on animals, plants, and insects in the past three decades.

A sudden widespread decline of frogs, with extinction of entire mountain-restricted species attributed to global warming [105]–[106], provided a dramatic awakening. There are multiple causes of the detailed processes involved in global amphibian declines and extinctions [107]–[108], but global warming is a key contributor and portends a planetary-scale mass extinction in the making unless action is taken to stabilize climate while also fighting biodiversity's other threats [109].

Mountain-restricted and polar-restricted species are particularly vulnerable. As isotherms move up the mountainside and poleward, so does the climate zone in which a given species can survive. If global warming continues unabated, many of these species will be effectively pushed off the planet. There are already reductions in the population and health of Arctic species in the southern parts of the Arctic, Antarctic species in the northern parts of the Antarctic, and alpine species worldwide [43].

A critical factor for survival of some Arctic species is retention of all-year sea ice. Continued growth of fossil fuel emissions will cause loss of all Arctic summer sea ice within several decades. In contrast, the scenario in Fig. 5A, with global warming peaking just over 1°C and then declining slowly, should allow summer sea ice to survive and then gradually increase to levels representative of recent decades.

The threat to species survival is not limited to mountain and polar species. Plant and animal distributions reflect the regional climates to which they are adapted. Although species attempt to migrate in response to climate change, their paths may be blocked by human-constructed obstacles or natural barriers such as coast lines and mountain ranges. As the shift of climate zones [110] becomes comparable to the range of some species, less mobile species can be driven to extinction. Because of extensive species interdependencies, this can lead to mass extinctions.

Rising sea level poses a threat to a large number of uniquely evolved endemic fauna living on islands in marine-dominated ecosystems, with those living on low lying islands being especially vulnerable. Evolutionary history on Bermuda offers numerous examples of the direct and indirect impact of changing sea level on evolutionary processes [111]–[112], with a number of taxa being extirpated due to habitat changes, greater competition, and island inundation [113]. Similarly, on Aldahabra Island in the Indian Ocean, land tortoises were exterminated during sea level high stands [114]. Vulnerabilities would be magnified by the speed of human-made climate change and the potentially large sea level rise [115].

IPCC [26] reviewed studies relevant to estimating eventual extinctions. They estimate that if global warming exceeds 1.6°C above preindustrial, 9–31 percent of species will be committed to extinction. With global warming of 2.9°C, an estimated 21–52 percent of species will be committed to extinction. A comprehensive study of biodiversity indicators over the past decade [116] reveals that, despite some local success in increasing extent of protected areas, overall indicators of pressures on biodiversity including that due to climate change are continuing to increase and indicators of the state of biodiversity are continuing to decline.

Mass extinctions occurred several times in Earth's history [117]–[118], often in conjunction with rapid climate change. New species evolved over millions of years, but those time scales are almost beyond human comprehension. If we drive many species to extinction we will leave a more desolate, monotonous planet for our children, grandchildren, and more generations than we can imagine. We will also undermine ecosystem functions (e.g., pollination which is critical for food production) and ecosystem resilience (when losing keystone species in food chains), as well as reduce functional diversity (critical for the ability of ecosystems to respond to shocks and stress) and genetic diversity that plays an important role for development of new medicines, materials, and sources of energy.

11.3.4 CORAL REEF ECOSYSTEMS

Coral reefs are the most biologically diverse marine ecosystem, often described as the rainforests of the ocean. Over a million species, most not yet described [119], are estimated to populate coral reef ecosystems generating crucial ecosystem services for at least 500 million people in tropical coastal areas. These ecosystems are highly vulnerable to the combined effects of ocean acidification and warming.

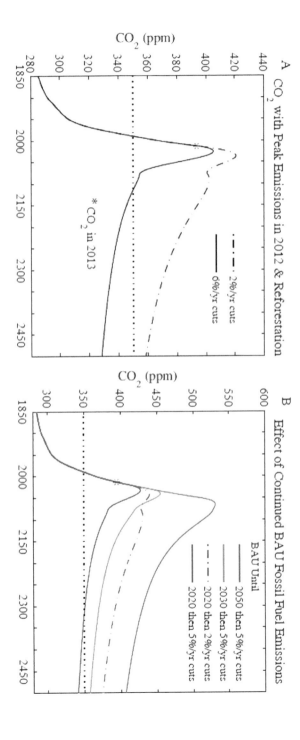

FIGURE 5: Atmospheric CO_2 if fossil fuel emissions reduced. (A) 6% or 2% annual cut begins in 2013 and 100 GtC reforestation drawdown occurs in 2031–2080. (B) effect of delaying onset of emission reduction.

Acidification arises as the ocean absorbs CO_2, producing carbonic acid [120], thus making the ocean more corrosive to the calcium carbonate shells (exoskeletons) of many marine organisms. Geochemical records show that ocean pH is already outside its range of the past several million years [121]–[122]. Warming causes coral bleaching, as overheated coral expel symbiotic algae and become vulnerable to disease and mortality [123]. Coral bleaching and slowing of coral calcification already are causing mass mortalities, increased coral disease, and reduced reef carbonate accretion, thus disrupting coral reef ecosystem health [40], [124].

Local human-made stresses add to the global warming and acidification effects, all of these driving a contraction of 1–2% per year in the abundance of reef-building corals [39]. Loss of the three-dimensional coral reef frameworks has consequences for all the species that depend on them. Loss of these frameworks also has consequences for the important roles that coral reefs play in supporting fisheries and protecting coastlines from wave stress. Consequences of lost coral reefs can be economically devastating for many nations, especially in combination with other impacts such as sea level rise and intensification of storms.

11.3.5 CLIMATE EXTREMES

Changes in the frequency and magnitude of climate extremes, of both moisture and temperature, are affected by climate trends as well as changing variability. Extremes of the hydrologic cycle are expected to intensify in a warmer world. A warmer atmosphere holds more moisture, so precipitation can be heavier and cause more extreme flooding. Higher temperatures, on the other hand, increase evaporation and can intensify droughts when they occur, as can expansion of the subtropics, as discussed above. Global models for the 21st century find an increased variability of precipitation minus evaporation [P-E] in most of the world, especially near the equator and at high latitudes [125]. Some models also show an intensification of droughts in the Sahel, driven by increasing greenhouse gases [126].

Observations of ocean salinity patterns for the past 50 years reveal an intensification of [P-E] patterns as predicted by models, but at an even faster rate. Precipitation observations over land show the expected general increase of precipitation poleward of the subtropics and decrease at lower latitudes [1], [26]. An increase of intense precipitation events has been found on much of the world's land area [127]–[129]. Evidence for widespread drought intensification is less clear and inherently difficult to confirm with available data because of the increase of time-integrated precipitation at most locations other than the subtropics. Data analyses have found an increase of drought intensity at many locations [130]–[131] The magnitude of change depends on the drought index employed [132], but soil moisture provides a good means to separate the effect of shifting seasonal precipitation and confirms an overall drought intensification [37].

Global warming of ~0.6°C since the 1970s (Fig. 3) has already caused a notable increase in the occurrence of extreme summer heat [46]. The likelihood of occurrence or the fractional area covered by 3-standard-deviation hot anomalies, relative to a base period (1951–1980) that was still within the range of Holocene climate, has increased by more than a factor of ten. Large areas around Moscow, the Mediterranean region, the United States and Australia have experienced such extreme anomalies in the past three years. Heat waves lasting for weeks have a devastating impact on human health: the European heat wave of summer 2003 caused over 70,000 excess deaths [133]. This heat record for Europe was surpassed already in 2010 [134]. The number of extreme heat waves has increased several-fold due to global warming [45]–[46], [135] and will increase further if temperatures continue to rise.

11.3.6 HUMAN HEALTH

Impacts of climate change cause widespread harm to human health, with children often suffering the most. Food shortages, polluted air, contaminated or scarce supplies of water, an expanding area of vectors causing infectious diseases, and more intensely allergenic plants are among the harmful impacts [26]. More extreme weather events cause physical and psychological harm. World health experts have concluded with "very high confidence" that climate change already contributes to the global burden of disease and premature death [26].

IPCC [26] projects the following trends, if global warming continue to increase, where only trends assigned very high confidence or high confidence are included: (i) increased malnutrition and consequent disorders, including those related to child growth and development, (ii) increased death, disease and injuries from heat waves, floods, storms, fires and droughts, (iii) increased cardio-respiratory morbidity and mortality associated with ground-level ozone. While IPCC also projects fewer deaths from cold, this positive effect is far outweighed by the negative ones.

Growing awareness of the consequences of human-caused climate change triggers anxiety and feelings of helplessness [136]–[137]. Children, already susceptible to age-related insecurities, face additional destabilizing insecurities from questions about how they will cope with future climate change [138]–[139]. Exposure to media ensures that children cannot escape hearing that their future and that of other species is at stake, and that the window of opportunity to avoid dramatic climate impacts is closing. The psychological health of our children is a priority, but denial of the truth exposes our children to even greater risk.

Health impacts of climate change are in addition to direct effects of air and water pollution. A clear illustration of direct effects of fossil fuels on human health was provided by an inadvertent experiment in China during the 1950–1980 period of central planning, when free coal for winter heating was provided to North China but not to the rest of the country. Analysis of the impact was made [140] using the most comprehensive data file ever compiled on mortality and air pollution in any developing country. A principal conclusion was that the 500 million residents of North China experienced during the 1990s a loss of more than 2.5 billion life years owing to the added air pollution, and an average reduction in life expectancy of 5.5 years. The degree of air pollution in China exceeded that in most of the world, yet assessments of total health effects must also include other fossil fuel caused air and water pollutants, as discussed in the following section on ecology and the environment.

The Text S1 has further discussion of health impacts of climate change.

11.3.7 ECOLOGY AND THE ENVIRONMENT

The ecological impact of fossil fuel mining increases as the largest, easiest to access, resources are depleted [141]. A constant fossil fuel production rate requires increasing energy input, but also use of more land, water, and diluents, with the production of more waste [142]. The increasing ecological and environmental impact of a given amount of useful fossil fuel energy is a relevant consideration in assessing alternative energy strategies.

Coal mining has progressively changed from predominantly underground mining to surface mining [143], including mountaintop removal with valley fill, which is now widespread in the Appalachian ecoregion in the United States. Forest cover and topsoil are removed, explosives are used to break up rocks to access coal, and the excess rock is pushed into adjacent valleys, where it buries existing streams. Burial of headwater streams causes loss of ecosystems that are important for nutrient cycling and production of organic matter for downstream food webs [144]. The surface alterations lead to greater storm runoff [145] with likely impact on downstream flooding. Water emerging from valley fills contain toxic solutes that have been linked to declines in watershed biodiversity [146]. Even with mine-site reclamation intended to restore pre-mined surface conditions, mine-derived chemical constituents are found in domestic well water [147]. Reclaimed areas, compared with unmined areas, are found to have increased soil density with decreased organic and nutrient content, and with reduced water infiltration rates [148]. Reclaimed areas have been found to produce little if any regrowth of woody vegetation even after 15 years [149], and, although this deficiency might be addressed via more effective reclamation methods, there remains a likely significant loss of carbon storage [149].

Oil mining has an increasing ecological footprint per unit delivered energy because of the decreasing size of new fields and their increased geographical dispersion; transit distances are greater and wells are deeper, thus requiring more energy input [145]. Useful quantitative measures of the increasing ecological impacts are provided by the history of oil development in Alberta, Canada for production of both conventional oil and tar sands development. The area of land required per barrel of produced oil increased by a factor of 12 between 1955 and 2006 [150] leading to ecosystem fragmentation by roads and pipelines needed to support the wells [151]. Additional escalation of the mining impact occurs as conventional oil mining is supplanted by tar sands development, with mining and land disturbance from the latter producing land use-related greenhouse gas emissions as much as 23 times greater than conventional oil production per unit area [152], but with substantial variability and uncertainty [152]–[153]. Much of the tar sands bitumen is extracted through surface mining that removes the "overburden" (i.e., boreal forest ecosystems) and tar sand from large areas to a depth up to 100 m, with ecological impacts downstream and in the mined area [154]. Although mined areas are supposed to be reclaimed, as in the case of mountaintop removal, there is no expectation that the ecological value of reclaimed areas will be equivalent to predevelopment condition [141], [155]. Landscape changes due to tar sands mining and reclamation cause a large loss of peatland and stored carbon, while also significantly reducing carbon sequestration potential [156]. Lake sediment cores document increased chemical pollution of ecosystems during the past several decades traceable to tar sands development [157] and snow and water samples indicate that recent levels of numerous pollutants exceeded local and national criteria for protection of aquatic organisms [158].

Gas mining by unconventional means has rapidly expanded in recent years, without commensurate understanding of the ecological, environmental and human health consequences [159]. The predominant approach is hydraulic fracturing ("fracking") of deep shale formations via injection of millions of gallons of water, sand and toxic chemicals under pressure, thus liberating methane [155], [160]. A large fraction of the injected water returns to the surface as wastewater containing high concentrations of heavy metals, oils, greases and soluble organic compounds [161]. Management of this wastewater is a major technical challenge, especially because the polluted waters can continue to backflow from the wells for many years [161]. Numerous instances of groundwater and river contamination have been cited [162]. High levels of methane leakage from fracking have been found [163], as well as nitrogen oxides and volatile organic compounds [159]. Methane leaks increase the climate impact of shale gas, but whether the leaks are sufficient to significantly alter the climate forcing by total natural gas development is uncertain [164]. Overall, environmental and ecologic threats posed by unconventional gas extraction are uncertain because of limited research, however evidence for groundwater pollution on both local and river basin scales is a major concern [165].

Today, with cumulative carbon emissions ~370 GtC from all fossil fuels, we are at a point of severely escalating ecological and environmental impacts from fossil fuel use and fossil fuel mining, as is apparent from the mountaintop removal for coal, tar sands extraction of oil, and fracking for gas. The ecological and environmental implications of scenarios with carbon emissions of 1000 GtC or greater, as discussed below, would be profound and should influence considerations of appropriate energy strategies.

11.3.8 SUMMARY: CLIMATE IMPACTS

Climate impacts accompanying global warming of 2°C or more would be highly deleterious. Already there are numerous indications of substantial effects in response to warming of the past few decades. That warming has brought global temperature close to if not slightly above the prior range of the Holocene. We conclude that an appropriate target would be to keep global temperature at a level within or close to the Holocene range. Global warming of 2°C would be well outside the Holocene range and far into the dangerous range.

11.4 TRANSIENT CLIMATE CHANGE

We must quantitatively relate fossil fuel emissions to global temperature in order to assess how rapidly fossil fuel emissions must be phased down to stay under a given temperature limit. Thus we must deal with both a transient carbon cycle and transient global climate change.

Global climate fluctuates stochastically and also responds to natural and human-made climate forcings [1], [166]. Forcings, measured in W/

m2 averaged over the globe, are imposed perturbations of Earth's energy balance caused by changing forcing agents such as solar irradiance and human-made greenhouse gases (GHGs). CO_2 accounts for more than 80% of the added GHG forcing in the past 15 years [64], [167] and, if fossil fuel emissions continue at a high level, CO_2 will be the dominant driver of future global temperature change.

We first define our method of calculating atmospheric CO_2 as a function of fossil fuel emissions. We then define our assumptions about the potential for drawing down atmospheric CO_2 via reforestation and increase of soil carbon, and we define fossil fuel emission reduction scenarios that we employ in our study. Finally we describe all forcings employed in our calculations of global temperature and the method used to simulate global temperature.

11.4.1 CARBON CYCLE AND ATMOSPHERIC CO_2

The carbon cycle defines the fate of CO_2 injected into the air by fossil fuel burning [1], [168] as the additional CO_2 distributes itself over time among surface carbon reservoirs: the atmosphere, ocean, soil, and biosphere. We use the dynamic-sink pulse-response function version of the well-tested Bern carbon cycle model [169], as described elsewhere [54], [170].

Specifically, we solve equations 3–6, 16–17, A.2.2, and A.3 of Joos et al. [169] using the same parameters and assumptions therein, except that initial (1850) atmospheric CO_2 is assumed to be 285.2 ppm [167]. Historical fossil fuel CO_2 emissions are from Boden et al. [5]. This Bern model incorporates non-linear ocean chemistry feedbacks and CO_2 fertilization of the terrestrial biosphere, but it omits climate-carbon feedbacks, e.g., assuming static global climate and ocean circulation. Therefore our results should be regarded as conservative, especially for scenarios with large emissions.

A pulse of CO_2 injected into the air decays by half in about 25 years as CO_2 is taken up by the ocean, biosphere and soil, but nearly one-fifth is still in the atmosphere after 500 years (Fig. 4A). Eventually, over hundreds of millennia, weathering of rocks will deposit all of this initial CO_2 pulse on the ocean floor as carbonate sediments [168].

Under equilibrium conditions a negative CO_2 pulse, i.e., artificial extraction and storage of some CO_2 amount, decays at about the same rate as a positive pulse (Fig. 4A). Thus if it is decided in the future that CO_2 must be extracted from the air and removed from the carbon cycle (e.g., by storing it underground or in carbonate bricks), the impact on atmospheric CO_2 amount will diminish in time. This occurs because carbon is exchanged among the surface carbon reservoirs as they move toward an equilibrium distribution, and thus, e.g., CO_2 out-gassing by the ocean can offset some of the artificial drawdown. The CO_2 extraction required to reach a given target atmospheric CO_2 level therefore depends on the prior emission history and target timeframe, but the amount that must be extracted substantially exceeds the net reduction of the atmospheric CO_2 level that will be achieved. We clarify this matter below by means of specific scenarios for capture of CO_2.

It is instructive to see how fast atmospheric CO_2 declines if fossil fuel emissions are instantly terminated (Fig. 4B). Halting emissions in 2015 causes CO_2 to decline to 350 ppm at century's end (Fig. 4B). A 20 year delay in halting emissions has CO_2 returning to 350 ppm at about 2300. With a 40 year delay, CO_2 does not return to 350 ppm until after 3000. These results show how difficult it is to get back to 350 ppm if emissions continue to grow for even a few decades.

These results emphasize the urgency of initiating emissions reduction [171]. As discussed above, keeping global climate close to the Holocene range requires a long-term atmospheric CO_2 level of about 350 ppm or less, with other climate forcings similar to today's levels. If emissions reduction had begun in 2005, reduction at 3.5%/year would have achieved 350 ppm at 2100. Now the requirement is at least 6%/year. Delay of emissions reductions until 2020 requires a reduction rate of 15%/year to achieve 350 ppm in 2100. If we assume only 50 GtC reforestation, and begin emissions reduction in 2013, the required reduction rate becomes about 9%/year.

11.4.2 REFORESTATION AND SOIL CARBON

Of course fossil fuel emissions will not suddenly terminate. Nevertheless, it is not impossible to return CO_2 to 350 ppm this century. Reforestation and increase of soil carbon can help draw down atmospheric CO_2. Fossil fuels account for ~80% of the CO_2 increase from preindustrial time, with land use/deforestation accounting for 20% [1], [170], [172]–[173]. Net deforestation to date is estimated to be 100 GtC (gigatons of carbon) with ±50% uncertainty [172].

Complete restoration of deforested areas is unrealistic, yet 100 GtC carbon drawdown is conceivable because: (1) the human-enhanced atmospheric CO_2 level increases carbon uptake by some vegetation and soils, (2) improved agricultural practices can convert agriculture from a CO_2 ource into a CO_2 sink [174], (3) biomass-burning power plants with CO_2 capture and storage can contribute to CO_2 drawdown.

Forest and soil storage of 100 GtC is challenging, but has other benefits. Reforestation has been successful in diverse places [175]. Minimum tillage with biological nutrient recycling, as opposed to plowing and chemical fertilizers, could sequester 0.4–1.2 GtC/year [176] while conserving water in soils, building agricultural resilience to climate change, and increasing productivity especially in smallholder rain-fed agriculture, thereby reducing expansion of agriculture into forested ecosystems [177]–[178]. Net tropical deforestation may have decreased in the past decade [179], but because of extensive deforestation in earlier decades [170], [172]–[173], [180]–[181] there is a large amount of land suitable for reforestation [182].

Use of bioenergy to draw down CO_2 should employ feedstocks from residues, wastes, and dedicated energy crops that do not compete with food crops, thus avoiding loss of natural ecosystems and cropland [183]–[185]. Reforestation competes with agricultural land use; land needs could decline by reducing use of animal products, as livestock now consume more than half of all crops [186].

Our reforestation scenarios assume that today's net deforestation rate (~1 GtC/year; see [54]) will stay constant until 2020, then linearly decrease to zero by 2030, followed by sinusoidal 100 GtC biospheric carbon storage over 2031–2080. Alternative timings do not alter conclusions about the potential to achieve a given CO_2 level such as 350 ppm.

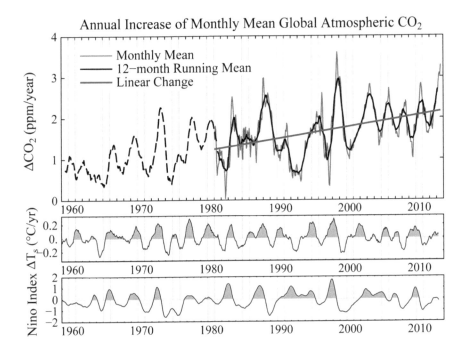

FIGURE 6: Annual increase of CO_2 based on data from the NOAA Earth System Research Laboratory [188]. Prior to 1981 the CO_2 change is based on only Mauna Loa, Hawaii. Temperature changes in lower diagram are 12-month running means for the globe and Niño3.4 area [16].

11.4.3 EMISSION REDUCTION SCENARIOS

A 6%/year decrease of fossil fuel emissions beginning in 2013, with 100 GtC reforestation, achieves a CO_2 decline to 350 ppm near the end of this century (Fig. 5A). Cumulative fossil fuel emissions in this scenario are ~129 GtC from 2013 to 2050, with an additional 14 GtC by 2100. If our assumed land use changes occur a decade earlier, CO_2 returns to 350 ppm several years earlier; however that has negligible effect on the maximum global temperature calculated below.

Delaying fossil fuel emission cuts until 2020 (with 2%/year emissions growth in 2012–2020) causes CO_2 to remain above 350 ppm (with associated impacts on climate) until 2300 (Fig. 5B). If reductions are delayed until 2030 or 2050, CO_2 remains above 350 ppm or 400 ppm, respectively, until well after 2500.

We conclude that it is urgent that large, long-term emission reductions begin soon. Even if a 6%/year reduction rate and 500 GtC are not achieved, it makes a huge difference when reductions begin. There is no practical justification for why emissions necessarily must even approach 1000 GtC.

11.4.4 CLIMATE FORCINGS

Atmospheric CO_2 and other GHGs have been well-measured for the past half century, allowing accurate calculation of their climate forcing. The growth rate of the GHG forcing has declined moderately since its peak values in the 1980s, as the growth rate of CH_4 and chlorofluorocarbons has slowed [187]. Annual changes of CO_2 are highly correlated with the El Niño cycle (Fig. 6). Two strong La Niñas in the past five years have depressed CO_2 growth as well as the global warming rate (Fig. 3). The CO_2 growth rate and warming rate can be expected to increase as we move into the next El Niño, with the CO_2 growth already reaching 3 ppm/year in mid-2013 [188]. The CO_2 climate forcing does not increase as rapidly as the CO_2 amount because of partial saturation of CO_2 absorption bands [75]. The GHG forcing is now increasing at a rate of almost 0.4 W/m^2 per decade [187].

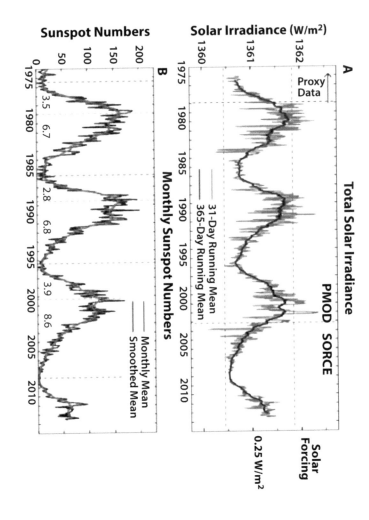

FIGURE 7: Solar irradiance and sunspot number in the era of satellite data (see text). Left scale is the energy passing through an area perpendicular to Sun-Earth line. Averaged over Earth's surface the absorbed solar energy is ~240 W/m², so the full amplitude of measured solar variability is ~0.25 W/m².

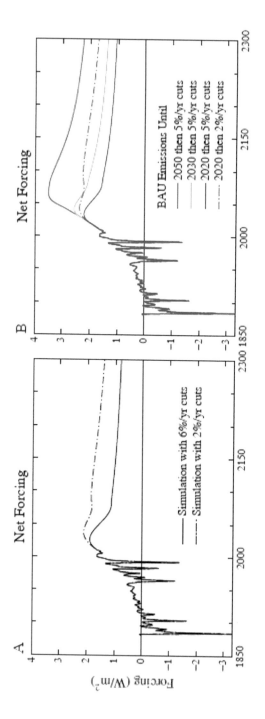

FIGURE 8: Climate forcings employed in our six main scenarios. Forcings through 2010 are as in [64].

Solar irradiance variations are sometimes assumed to be the most likely natural driver of climate change. Solar irradiance has been measured from satellites since the late 1970s (Fig. 7). These data are from a composite of several satellite-measured time series. Data through 28 February 2003 are from [189] and Physikalisch Meteorologisches Observatorium Davos, World Radiation Center. Subsequent update is from University of Colorado Solar Radiation & Climate Experiment (SORCE). Data sets are concatenated by matching the means over the first 12 months of SORCE data. Monthly sunspot numbers (Fig. 7) support the conclusion that the solar irradiance in the current solar cycle is significantly lower than in the three preceding solar cycles. Amplification of the direct solar forcing is conceivable, e.g., through effects on ozone or atmospheric condensation nuclei, but empirical data place a factor of two upper limit on the amplification, with the most likely forcing in the range 100–120% of the directly measured solar irradiance change [64].

Recent reduced solar irradiance (Fig. 7) may have decreased the forcing over the past decade by about half of the full amplitude of measured irradiance variability, thus yielding a negative forcing of, say, -0.12 W/m^2. This compares with a decadal increase of the GHG forcing that is positive and about three times larger in magnitude. Thus the solar forcing is not negligible and might partially account for the slowdown in global warming in the past decade [17]. However, we must (1) compare the solar forcing with the net of other forcings, which enhances the importance of solar change, because the net forcing is smaller than the GHG forcing, and (2) consider forcing changes on longer time scales, which greatly diminishes the importance of solar change, because solar variability is mainly oscillatory.

Human-made tropospheric aerosols, which arise largely from fossil fuel use, cause a substantial negative forcing. As noted above, two independent analyses [64], [72] yield a total (direct plus indirect) aerosol forcing in the past decade of about -1.5 W/m^2, half the magnitude of the GHG forcing and opposite in sign. That empirical aerosol forcing assessment for the past decade is consistent with the climate forcings scenario (Fig. 8) that we use for the past century in the present and prior studies [64], [190]. Supplementary Table S1 specifies the historical forcings and Table S2 gives several scenarios for future forcings.

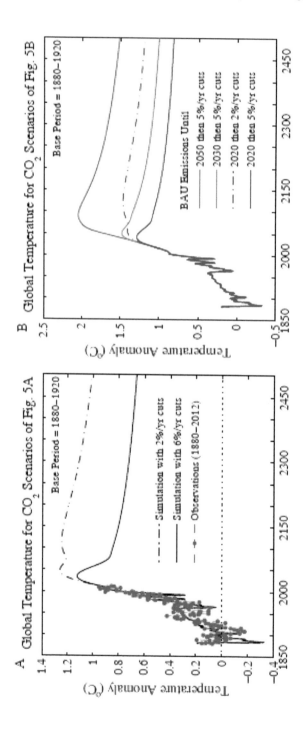

FIGURE 9: Simulated global temperature relative to 1880–1920 mean for CO_2 scenarios of Figure 5.

11.4.5 FUTURE CLIMATE FORCINGS

Future global temperature change should depend mainly on atmospheric CO_2, at least if fossil fuel emissions remain high. Thus to provide the clearest picture of the CO_2 effect, we approximate the net future change of human-made non-CO_2 forcings as zero and we exclude future changes of natural climate forcings, such as solar irradiance and volcanic aerosols. Here we discuss possible effects of these approximations.

Uncertainties in non-CO_2 forcings concern principally solar, aerosol and other GHG forcings. Judging from the sunspot numbers (Fig. 7B and [191]) for the past four centuries, the current solar cycle is almost as weak as the Dalton Minimum of the late 18th century. Conceivably irradiance could decline further to the level of the Maunder Minimum of the late 17th century [192]–[193]. For our simulation we choose an intermediate path between recovery to the level before the current solar cycle and decline to a still lower level. Specifically, we keep solar irradiance fixed at the reduced level of 2010, which is probably not too far off in either direction. Irradiance in 2010 is about 0.1 W/m^2 less than the mean of the prior three solar cycles, a decrease of forcing that would be restored by the CO_2 increase within 3–4 years at its current growth rate. Extensive simulations [17], [194] confirm that the effect of solar variability is small compared with GHGs if CO_2 emissions continue at a high level. However, solar forcing can affect the magnitude and detection of near-term warming. Also, if rapidly declining GHG emissions are achieved, changes of solar forcing will become relatively more important.

Aerosols present a larger uncertainty. Expectations of decreases in large source regions such as China [195] may be counteracted by aerosol increases other places as global population continues to increase. Our assumption of unchanging human-made aerosols could be substantially off in either direction. For the sake of interpreting on-going and future climate change it is highly desirable to obtain precise monitoring of the global aerosol forcing [73].

Non-CO_2 GHG forcing has continued to increase at a slow rate since 1995 (Fig. 6 in [64]). A desire to constrain climate change may help reduce emissions of these gases in the future. However, it will be difficult to prevent or fully offset positive forcing from increasing N_2O, as its largest source is associated with food production and the world's population is continuing to rise.

On the other hand, we are also probably underestimating a negative aerosol forcing, e.g., because we have not included future volcanic aerosols. Given the absence of large volcanic eruptions in the past two decades (the last one being Mount Pinatubo in 1991), multiple volcanic eruptions would cause a cooling tendency [196] and reduce heat storage in the ocean [197].

Overall, we expect the errors due to our simple approximation of non-$C_2$2 forcings to be partially off-setting. Specifically, we have likely underestimated a positive forcing by non-CO_2 GHGs, while also likely underestimating a negative aerosol forcing.

Note that uncertainty in forcings is partly obviated via the focus on Earth's energy imbalance in our analysis. The planet's energy imbalance is an integrative quantity that is especially useful for a case in which some of the forcings are uncertain or unmeasured. Earth's measured energy imbalance includes the effects of all forcings, whether they are measured or not.

11.4.6 SIMULATIONS OF FUTURE GLOBAL TEMPERATURE

We calculate global temperature change for a given CO2 scenario using a climate response function (Table S3) that accurately replicates results from a global climate model with sensitivity 3°C for doubled CO2 [64]. A best estimate of climate sensitivity close to 3°C for doubled CO2 has been inferred from paleoclimate data [51]–[52]. This empirical climate sensitivity is generally consistent with that of global climate models [1], but the empirical approach makes the inferred high sensitivity more certain and the quantitative evaluation more precise. Because this climate sensitivity is derived from empirical data on how Earth responded to past changes of boundary conditions, including atmospheric composition, our conclusions about limits on fossil fuel emissions can be regarded as largely independent of climate models. The detailed temporal and geographical response of the climate system to the rapid human-made change of climate forcings is not well-constrained by empirical data, because there is no faithful paleoclimate analog. Thus climate models necessarily play an important role in assessing practical implications of climate change. Nevertheless, it is possible to draw important conclusions with transparent computations. A simple response function (Green's function) calculation [64] yields an estimate of global mean temperature change in response to a specified time series for global climate forcing. This approach accounts for the delayed response of the climate system caused by the large thermal inertia of the ocean, yielding a global mean temporal response in close accord with that obtained from global climate models.

Tables S1 and S2 in Supporting Information give the forcings we employ and Table S3 gives the climate response function for our Green's function calculation, defined by equation 2 of [64]. The Green's function is driven by the net forcing, which, with the response function, is sufficient information for our results to be reproduced. However, we also include the individual forcings in Table S1, in case researchers wish to replace specific forcings or use them for other purposes.

Simulated global temperature (Fig. 9) is for CO_2 scenarios of Fig. 5. Peak global warming is ~1.1°C, declining to less than 1°C by mid-century, if CO_2 emissions are reduced 6%/year beginning in 2013. In contrast, warming reaches 1.5°C and stays above 1°C until after 2400 if emissions continue to increase until 2030, even though fossil fuel emissions are phased out rapidly (5%/year) after 2030 and 100 GtC reforestation occurs during 2030–2080. If emissions continue to increase until 2050, simulated warming exceeds 2°C well into the 22nd century.

Increased global temperature persists for many centuries after the climate forcing declines, because of the thermal inertia of the ocean [198]. Some temperature reduction is possible if the climate forcing is reduced rapidly, before heat has penetrated into the deeper ocean. Cooling by a few tenths of a degree in Fig. 9 is a result mainly of the 100 GtC biospheric uptake of CO_2 during 2030–2080. Note the longevity of the warming, especially if emissions reduction is as slow as 2%/year, which might be considered to be a rapid rate of reduction.

The temporal response of the real world to the human-made climate forcing could be more complex than suggested by a simple response function calculation, especially if rapid emissions growth continues, yielding an unprecedented climate forcing scenario. For example, if ice sheet mass loss becomes rapid, it is conceivable that the cold fresh water added to the ocean could cause regional surface cooling [199], perhaps even at a point when sea level rise has only reached a level of the order of a meter [200]. However, any uncertainty in the surface thermal response this century due to such phenomena has little effect on our estimate of the dangerous level of emissions. The long lifetime of the fossil fuel carbon in the climate system and the persistence of ocean warming for millennia [201] provide sufficient time for the climate system to achieve full response to the fast feedback processes included in the 3°C climate sensitivity.

Indeed, the long lifetime of fossil fuel carbon in the climate system and persistence of the ocean warming ensure that "slow" feedbacks, such as ice sheet disintegration, changes of the global vegetation distribution, melting of permafrost, and possible release of methane from methane hydrates on continental shelves, would also have time to come into play. Given the unprecedented rapidity of the human-made climate forcing, it is difficult to establish how soon slow feedbacks will become important, but clearly slow feedbacks should be considered in assessing the "dangerous" level of global warming, as discussed in the next section.

11.5 DANGER OF INITIATING UNCONTROLLABLE CLIMATE CHANGE

Our calculated global warming as a function of CO_2 amount is based on equilibrium climate sensitivity 3°C for doubled CO_2. That is the central climate sensitivity estimate from climate models [1], and it is consistent with climate sensitivity inferred from Earth's climate history [51]–[52]. However, this climate sensitivity includes only the effects of fast feedbacks of the climate system, such as water vapor, clouds, aerosols, and sea ice. Slow feedbacks, such as change of ice sheet area and climate-driven changes of greenhouse gases, are not included.

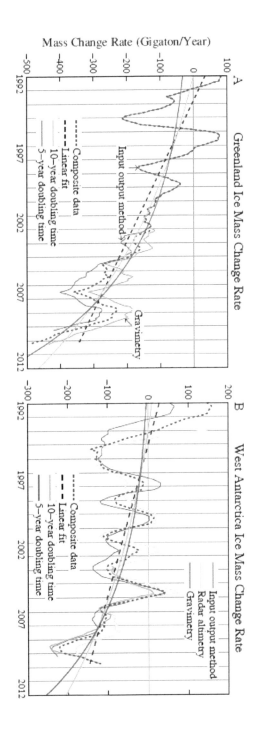

FIGURE 10: Annual Greenland and West Antarctic ice mass changes as estimated via alternative methods. Data were read from Figure 4 of Shepherd et al. [23] and averaged over the available records.

11.5.1 SLOW CLIMATE FEEDBACKS
AND IRREVERSIBLE CLIMATE CHANGE

Excluding slow feedbacks was appropriate for simulations of the past century, because we know the ice sheets were stable then and our climate simulations used observed greenhouse gas amounts that included any contribution from slow feedbacks. However, we must include slow feedbacks in projections of warming for the 21st century and beyond. Slow feedbacks are important because they affect climate sensitivity and because their instigation is related to the danger of passing "points of no return", beyond which irreversible consequences become inevitable, out of humanity's control.

Antarctic and Greenland ice sheets present the danger of change with consequences that are irreversible on time scales important to society [1]. These ice sheets required millennia to grow to their present sizes. If ice sheet disintegration reaches a point such that the dynamics and momentum of the process take over, at that point reducing greenhouse gases may be unable to prevent major ice sheet mass loss, sea level rise of many meters, and worldwide loss of coastal cities – a consequence that is irreversible for practical purposes. Interactions between the ocean and ice sheets are particularly important in determining ice sheet changes, as a warming ocean can melt the ice shelves, the tongues of ice that extend from the ice sheets into the ocean and buttress the large land-based ice sheets [92], [202]–[203]. Paleoclimate data for sea level change indicate that sea level changed at rates of the order of a meter per century [81]–[83], even at times when the forcings driving climate change were far weaker than the human-made forcing. Thus, because ocean warming is persistent for centuries, there is a danger that large irreversible change could be initiated by excessive ocean warming.

Paleoclimate data are not as helpful for defining the likely rate of sea level rise in coming decades, because there is no known case of growth of a positive (warming) climate forcing as rapid as the anthropogenic change. The potential for unstable ice sheet disintegration is controversial, with opinion varying from likely stability of even the (marine) West Antarctic ice sheet [94] to likely rapid non-linear response extending up to multi-meter sea level rise [97]–[98]. Data for the modern rate of annual ice sheet mass changes indicate an accelerating rate of mass loss consistent with a mass loss doubling time of a decade or less (Fig. 10). However, we do not know the functional form of ice sheet response to a large persistent climate forcing. Longer records are needed for empirical assessment of this ostensibly nonlinear behavior.

Greenhouse gas amounts in the atmosphere, most importantly CO_2 and CH_4, change in response to climate change, i.e., as a feedback, in addition to the immediate gas changes from human-caused emissions. As the ocean warms, for example, it releases CO_2 to the atmosphere, with one principal mechanism being the simple fact that the solubility of CO_2 decreases as the water temperature rises [204]. We also include in the category of slow feedbacks the global warming spikes, or "hyperthermals", that have occurred a number of times in Earth's history during the course of slower global warming trends. The mechanisms behind these hyperthermals are poorly understood, as discussed below, but they are characterized by the injection into the surface climate system of a large amount of carbon in the form of CH_4 and/or CO_2 on the time scale of a millennium [205]–[207]. The average rate of injection of carbon into the climate system during these hyperthermals was slower than the present human-made injection of fossil fuel carbon, yet it was faster than the time scale for removal of carbon from the surface reservoirs via the weathering process [3], [208], which is tens to hundreds of thousands of years.

Methane hydrates—methane molecules trapped in frozen water molecule cages in tundra and on continental shelves—and organic matter such as peat locked in frozen soils (permafrost) are likely mechanisms in the past hyperthermals, and they provide another climate feedback with the potential to amplify global warming if large scale thawing occurs [209]–[210]. Paleoclimate data reveal instances of rapid global warming, as much as 5–6°C, as a sudden additional warming spike during a longer period of gradual warming [see Text S1]. The candidates for the carbon injected into the climate system during those warmings are methane hydrates on continental shelves destabilized by sea floor warming [211] and carbon released from frozen soils [212]. As for the present, there are reports of methane release from thawing permafrost on land [213] and from sea-bed methane hydrate deposits [214], but amounts so far are small and the data are snapshots that do not prove that there is as yet a temporal increase of emissions.

There is a possibility of rapid methane hydrate or permafrost emissions in response to warming, but that risk is largely unquantified [215]. The time needed to destabilize large methane hydrate deposits in deep sediments is likely millennia [215]. Smaller but still large methane hydrate amounts below shallow waters as in the Arctic Ocean are more vulnerable; the methane may oxidize to CO_2 in the water, but it will still add to the long-term burden of CO_2 in the carbon cycle. Terrestrial permafrost emissions of CH_4 and CO_2 likely can occur on a time scale of a few decades to several centuries if global warming continues [215]. These time scales are within the lifetime of anthropogenic CO_2, and thus these feedbacks must be considered in estimating the dangerous level of global warming. Because human-made warming is more rapid than natural long-term warmings in the past, there is concern that methane hydrate or peat feedbacks could be more rapid than the feedbacks that exist in the paleoclimate record.

Climate model studies and empirical analyses of paleoclimate data can provide estimates of the amplification of climate sensitivity caused by slow feedbacks, excluding the singular mechanisms that caused the hyperthermal events. Model studies for climate change between the Holocene and the Pliocene, when Earth was about 3°C warmer, find that slow feedbacks due to changes of ice sheets and vegetation cover amplified the fast feedback climate response by 30–50% [216]. These same slow feedbacks are estimated to amplify climate sensitivity by almost a factor of two for the climate change between the Holocene and the nearly ice-free climate state that existed 35 million years ago [54].

11.5.2 IMPLICATION FOR CARBON EMISSIONS TARGET

Evidence presented under Climate Impacts above makes clear that 2°C global warming would have consequences that can be described as disastrous. Multiple studies [12], [198], [201] show that the warming would be very long lasting. The paleoclimate record and changes underway in the Arctic and on the Greenland and Antarctic ice sheets with only today's warming imply that sea level rise of several meters could be expected. Increased climate extremes, already apparent at 0.8°C warming [46], would be more severe. Coral reefs and associated species, already stressed with current conditions [40], would be decimated by increased acidification, temperature and sea level rise. More generally, humanity and nature, the modern world as we know it, is adapted to the Holocene climate that has existed more than 10,000 years. Warming of 1°C relative to 1880–1920 keeps global temperature close to the Holocene range, but warming of 2°C, to at least the Eemian level, could cause major dislocations for civilization.

However, distinctions between pathways aimed at ~1°C and 2°C warming are much greater and more fundamental than the numbers 1°C and 2°C themselves might suggest. These fundamental distinctions make scenarios with 2°C or more global warming far more dangerous; so dangerous, we suggest, that aiming for the 2°C pathway would be foolhardy.

First, most climate simulations, including ours above and those of IPCC [1], do not include slow feedbacks such as reduction of ice sheet size with global warming or release of greenhouse gases from thawing tundra. These exclusions are reasonable for a ~1°C scenario, because global temperature barely rises out of the Holocene range and then begins to subside. In contrast, global warming of 2°C or more is likely to bring slow feedbacks into play. Indeed, it is slow feedbacks that cause long-term climate sensitivity to be high in the empirical paleoclimate record [51]–[52]. The lifetime of fossil fuel CO_2 in the climate system is so long that it must be assumed that these slow feedbacks will occur if temperature rises well above the Holocene range.

Second, scenarios with 2°C or more warming necessarily imply expansion of fossil fuels into sources that are harder to get at, requiring greater energy using extraction techniques that are increasingly invasive, destructive and polluting. Fossil fuel emissions through 2012 total ~370 GtC (Fig. 2). If subsequent emissions decrease 6%/year, additional emissions are ~130 GtC, for a total ~500 GtC fossil fuel emissions. This 130 GtC can be obtained mainly from the easily extracted conventional oil and gas reserves (Fig. 2), with coal use rapidly phased out and unconventional fossil fuels left in the ground. In contrast, 2°C scenarios have total emissions of the order of 1000 GtC. The required additional fossil fuels will involve exploitation of tar sands, tar shale, hydrofracking for oil and gas, coal mining, drilling in the Arctic, Amazon, deep ocean, and other remote regions, and possibly exploitation of methane hydrates. Thus 2°C scenarios result in more CO_2 per unit useable energy, release of substantial CH_4 via the mining process and gas transportation, and release of CO_2 and other gases via destruction of forest "overburden" to extract subterranean fossil fuels.

Third, with our ~1°C scenario it is more likely that the biosphere and soil will be able to sequester a substantial portion of the anthropogenic fossil fuel CO_2 carbon than in the case of 2°C or more global warming. Empirical data for the CO_2 "airborne fraction", the ratio of observed atmospheric CO_2 increase divided by fossil fuel CO_2 emissions, show that almost half of the emissions is being taken up by surface (terrestrial and ocean) carbon reservoirs [187], despite a substantial but poorly measured contribution of anthropogenic land use (deforestation and agriculture) to airborne CO_2 [179], [216]. Indeed, uptake of CO_2 by surface reservoirs has at least kept pace with the rapid growth of emissions [187]. Increased uptake in the past decade may be a consequence of a reduced rate of deforestation [217] and fertilization of the biosphere by atmospheric CO_2 and nitrogen deposition [187]. With the stable climate of the ~1°C scenario it is plausible that major efforts in reforestation and improved agricultural practices [15], [173], [175]–[177], with appropriate support provided to developing countries, could take up an amount of carbon comparable to the 100 GtC in our ~1°C scenario. On the other hand, with warming of 2°C or more, carbon cycle feedbacks are expected to lead to substantial additional atmospheric CO_2 [218]–[219], perhaps even making the Amazon rainforest a source of CO_2 [219]–[220].

Fourth, a scenario that slows and then reverses global warming makes it possible to reduce other greenhouse gases by reducing their sources [75], [221]. The most important of these gases is CH_4, whose reduction in turn reduces tropospheric O_3 and stratospheric H_2O. In contrast, chemistry modeling and paleoclimate records [222] show that trace gases increase with global warming, making it unlikely that overall atmospheric CH_4 will decrease even if a decrease is achieved in anthropogenic CH_4 sources. Reduction of the amount of atmospheric CH_4 and related gases is needed to counterbalance expected forcing from increasing N_2O and decreasing sulfate aerosols.

Now let us compare the 1°C (500 GtC fossil fuel emissions) and the 2°C (1000 GtC fossil fuel emissions) scenarios. Global temperature in 2100 would be close to 1°C in the 500 GtC scenario, and it is less than

1°C if 100 GtC uptake of carbon by the biosphere and soil is achieved via improved agricultural and forestry practices (Fig. 9). In contrast, the 1000 GtC scenario, although nominally designed to yield a fast-feedback climate response of ~ 2°C, would yield a larger eventual warming because of slow feedbacks, probably at least 3°C.

11.5.3 DANGER OF UNCONTROLLABLE CONSEQUENCES

Inertia of the climate system reduces the near-term impact of human-made climate forcings, but that inertia is not necessarily our friend. One implication of the inertia is that climate impacts "in the pipeline" may be much greater than the impacts that we presently observe. Slow climate feedbacks add further danger of climate change running out of humanity's control. The response time of these slow feedbacks is uncertain, but there is evidence that some of these feedbacks already are underway, at least to a minor degree. Paleoclimate data show that on century and millennial time scales the slow feedbacks are predominately amplifying feedbacks.

The inertia of energy system infrastructure, i.e., the time required to replace fossil fuel energy systems, will make it exceedingly difficult to avoid a level of atmospheric CO_2 that would eventually have highly undesirable consequences. The danger of uncontrollable and irreversible consequences necessarily raises the question of whether it is feasible to extract CO_2 from the atmosphere on a large enough scale to affect climate change.

11.6 CARBON EXTRACTION

We have shown that extraordinarily rapid emission reductions are needed to stay close to the 1°C scenario. In absence of extraordinary actions, it is likely that growing climate disruptions will lead to a surge of interest in "geo-engineering" designed to minimize human-made climate change [223]. Such efforts must remove atmospheric CO_2, if they are to address direct CO_2 effects such as ocean acidification as well as climate change. Schemes such as adding sulfuric acid aerosols to the stratosphere to reflect sunlight [224], an attempt to mask one pollutant with another, is a temporary band-aid for a problem that will last for millennia; besides it fails to address ocean acidification and may have other unintended consequences [225].

11.6.1 POTENTIAL FOR CARBON EXTRACTION

At present there are no proven technologies capable of large-scale air capture of CO_2. It has been suggested that, with strong research and development support and industrial scale pilot projects sustained over decades, costs as low as ~\$500/tC may be achievable [226]. Thermodynamic constraints [227] suggest that this cost estimate may be low. An assessment by the American Physical Society [228] argues that the lowest currently achievable cost, using existing approaches, is much greater (\$600/tCO_2 or \$2200/tC).

The cost of capturing 50 ppm of CO_2, at \$500/tC (~\$135/tCO_2), is ~\$50 trillion (1 ppm CO_2 is ~2.12 GtC), but more than \$200 trillion for the price estimate of the American Physical Society study. Moreover, the resulting atmospheric CO_2 reduction will ultimately be less than 50 ppm for the reasons discussed above. For example, let us consider the scenario of Fig. 5B in which emissions continue to increase until 2030 before decreasing at 5%/year – this scenario yields atmospheric CO_2 of 410 ppm in 2100. Using our carbon cycle model we calculate that if we extract 100 ppm of CO_2 from the air over the period 2030–2100 (10/7 ppm per year), say storing that CO_2 in carbonate bricks, the atmospheric CO_2 amount in 2100 will be reduced 52 ppm to 358 ppm, i.e., the reduction of airborne CO_2 is about half of the amount extracted from the air and stored. The estimated cost of this 52 ppm CO_2 reduction is \$100–400 trillion.

The cost of CO_2 capture and storage conceivably may decline in the future. Yet the practicality of carrying out such a program with alacrity in response to a climate emergency is dubious. Thus it may be appropriate to add a CO_2 removal cost to the current price of fossil fuels, which would both reduce ongoing emissions and provide resources for future cleanup.

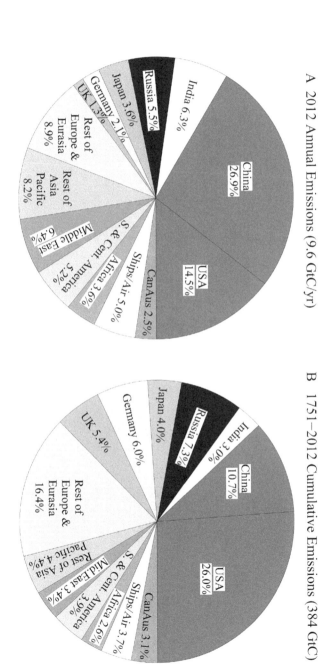

FIGURE 11: Fossil fuel CO$_2$ emissions. (A) 2012 emissions by source region, and (B) cumulative 1751–2012 emissions. Results are an update of Fig. 10 of [190] using data from [5].

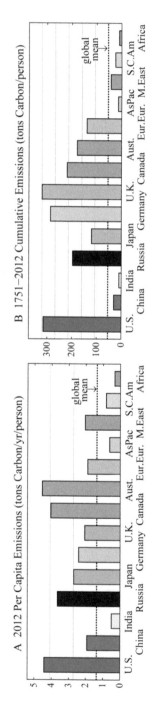

FIGURE 12: Per capita fossil fuel CO_2 emissions. Countries, regions and data sources are the same as in Fig. 11. Horizontal lines are the global mean and multiples of the global mean.

11.6.2 RESPONSIBILITY FOR CARBON EXTRACTION

We focus on fossil fuel carbon, because of its long lifetime in the carbon cycle. Reversing the effects of deforestation is also important and there will need to be incentives to achieve increased carbon storage in the biosphere and soil, but the crucial requirement now is to limit the amount of fossil fuel carbon in the air.

The high cost of carbon extraction naturally raises the question of responsibility for excess fossil fuel CO_2 in the air. China has the largest CO_2 emissions today (Fig. 11A), but the global warming effect is closely proportional to cumulative emissions [190]. The United States is responsible for about one-quarter of cumulative emissions, with China next at about 10% (Fig. 11B). Cumulative responsibilities change rather slowly (compare Fig. 10 of 190). Estimated per capita emissions (Fig. 12) are based on population estimates for 2009–2011.

Various formulae might be devised to assign costs of CO_2 air capture, should removal prove essential for maintaining acceptable climate. For the sake of estimating the potential cost, let us assume that it proves necessary to extract 100 ppm of CO_2 (yielding a reduction of airborne CO_2 of about 50 ppm) and let us assign each country the responsibility to clean up its fraction of cumulative emissions. Assuming a cost of $500/tC (~$135/ tCO_2) yields a cost of $28 trillion for the United States, about $90,000 per individual. Costs would be slightly higher for a UK citizen, but less for other nations (Fig. 12B).

Cost of CO_2 capture might decline, but the cost estimate used is more than a factor of four smaller than estimated by the American Physical Society [228] and 50 ppm is only a moderate reduction. The cost should also include safe permanent disposal of the captured CO_2, which is a substantial mass. For the sake of scaling the task, note that one GtC, made into carbonate bricks, would produce the volume of ~3000 Empire State buildings or ~1200 Great Pyramids of Giza. Thus the 26 ppm assigned to the United States, if made into carbonate bricks, would be equivalent to the stone in 165,000 Empire State buildings or 66,000 Great Pyramids of Giza. This is not intended as a practical suggestion: carbonate bricks are not a good building material, and the transport and construction costs would be additional.

The point of this graphic detail is to make clear the magnitude of the cleanup task and potential costs, if fossil fuel emissions continue unabated. More useful and economic ways of removing CO_2 may be devised with the incentive of a sufficient carbon price. For example, a stream of pure CO_2 becomes available for capture and storage if biomass is used as the fuel for power plants or as feedstock for production of liquid hydrocarbon fuels. Such clean energy schemes and improved agricultural and forestry practices are likely to be more economic than direct air capture of CO_2, but they must be carefully designed to minimize undesirable impacts and the amount of CO_2 that can be extracted on the time scale of decades will be limited, thus emphasizing the need to limit the magnitude of the cleanup task.

11.7 POLICY IMPLICATIONS

Human-made climate change concerns physical sciences, but leads to implications for policy and politics. Conclusions from the physical sciences, such as the rapidity with which emissions must be reduced to avoid obviously unacceptable consequences and the long lag between emissions and consequences, lead to implications in social sciences, including economics, law and ethics. Intergovernmental climate assessments [1], [14] purposely are not policy prescriptive. Yet there is also merit in analysis and discussion of the full topic through the objective lens of science, i.e., "connecting the dots" all the way to policy implications.

11.7.1 ENERGY AND CARBON PATHWAYS: A FORK IN THE ROAD

The industrial revolution began with wood being replaced by coal as the primary energy source. Coal provided more concentrated energy, and thus was more mobile and effective. We show data for the United States (Fig. 13) because of the availability of a long data record that includes wood [229]. More limited global records yield a similar picture [Fig. 14], the largest difference being global coal now at ~30% compared with ~20% in the United States. Economic progress and wealth generation were further spurred in the twentieth century by expansion into liquid and gaseous fossil fuels, oil and gas being transported and burned more readily than coal. Only in the latter part of the twentieth century did it become clear that long-lived combustion products from fossil fuels posed a global climate threat, as formally acknowledged in the 1992 Framework Convention on Climate Change [6]. However, efforts to slow emissions of the principal atmospheric gas driving climate change, CO_2, have been ineffectual so far (Fig. 1).

Consequently, at present, as the most easily extracted oil and gas reserves are being depleted, we stand at a fork in the road to our energy and carbon future. Will we now feed our energy needs by pursuing difficult to extract fossil fuels, or will we pursue energy policies that phase out carbon emissions, moving on to the post fossil fuel era as rapidly as practical?

This is not the first fork encountered. Most nations agreed to the Framework Convention on Climate Change in 1992 [6]. Imagine if a bloc of countries favoring action had agreed on a common gradually rising carbon fee collected within each of country at domestic mines and ports of entry. Such nations might place equivalent border duties on products from nations not having a carbon fee and they could rebate fees to their domestic industry for export products to nations without an equivalent carbon fee. The legality of such a border tax adjustment under international trade law is untested, but is considered to be plausibly consistent with trade principles [230]. As the carbon fee gradually rose and as additional nations, for their own benefit, joined this bloc of nations, development of carbon-free energies and energy efficiency would have been spurred. If the carbon fee had begun in 1995, we calculate that global emissions would have needed to decline 2.1%/year to limit cumulative fossil fuel emissions to 500 GtC. A start date of 2005 would have required a reduction of 3.5%/year for the same result.

The task faced today is more difficult. Emissions reduction of 6%/year and 100 GtC storage in the biosphere and soils are needed to get CO2 back to 350 ppm, the approximate requirement for restoring the planet's energy balance and stabilizing climate this century. Such a pathway is exceedingly difficult to achieve, given the current widespread absence of policies to drive rapid movement to carbon-free energies and the lifetime of energy infrastructure in place.

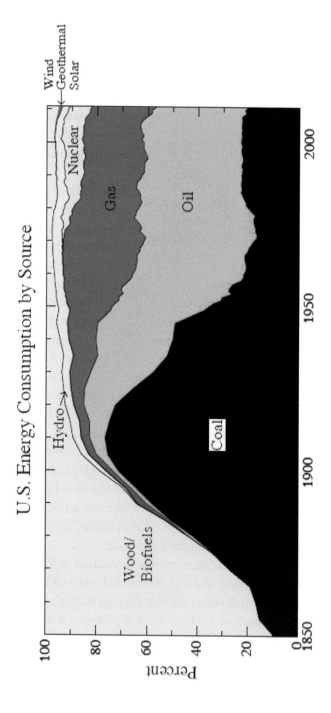

FIGURE 13: United States energy consumption [229].

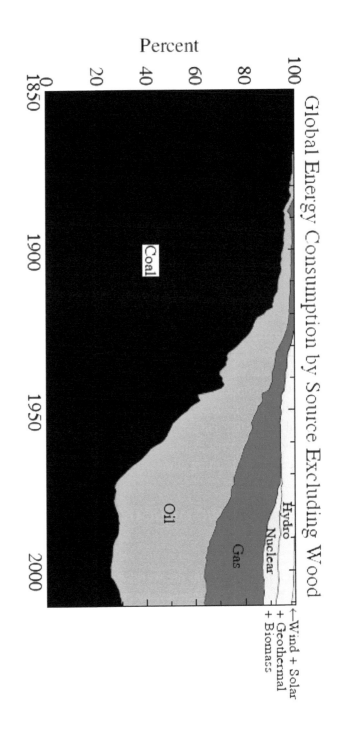

FIGURE 14: World energy consumption for indicated fuels, which excludes wood [4].

Yet we suggest that a pathway is still conceivable that could restore planetary energy balance on the century time scale. That path requires policies that spur technology development and provide economic incentives for consumers and businesses such that social tipping points are reached where consumers move rapidly to energy conservation and low carbon energies. Moderate overshoot of required atmospheric CO_2 levels can possibly be counteracted via incentives for actions that more-or-less naturally sequester carbon. Developed countries, responsible for most of the excess CO_2 in the air, might finance extensive efforts in developing countries to sequester carbon in the soil and in forest regrowth on marginal lands as described above. Burning sustainably designed biofuels in power plants, with the CO_2 captured and sequestered, would also help draw down atmospheric CO_2. This pathway would need to be taken soon, as the magnitude of such carbon extractions is likely limited and thus not a solution to unfettered fossil fuel use.

The alternative pathway, which the world seems to be on now, is continued extraction of all fossil fuels, including development of unconventional fossil fuels such as tar sands, tar shale, hydrofracking to extract oil and gas, and exploitation of methane hydrates. If that path (with 2%/year growth) continues for 20 years and is then followed by 3%/year emission reduction from 2033 to 2150, we find that fossil fuel emissions in 2150 would total 1022 GtC. Extraction of the excess CO_2 from the air in this case would be very expensive and perhaps implausible, and warming of the ocean and resulting climate impacts would be practically irreversible.

11.7.2 ECONOMIC IMPLICATIONS: NEED FOR A CARBON FEE

The implication is that the world must move rapidly to carbon-free energies and energy efficiency, leaving most remaining fossil fuels in the ground, if climate is to be kept close to the Holocene range and climate disasters averted. Is rapid change possible?

The potential for rapid change can be shown by examples. A basic requirement for phasing down fossil fuel emissions is abundant carbon-free electricity, which is the most rapidly growing form of energy and also has the potential to provide energy for transportation and heating of buildings. In one decade (1977–1987), France increased its nuclear power production 15-fold, with the nuclear portion of its electricity increasing from 8% to 70% [231]. In one decade (2001–2011) Germany increased the non-hydroelectric renewable energy portion of its electricity from 4% to 19%, with fossil fuels decreasing from 63% to 61% (hydroelectric decreased from 4% to 3% and nuclear power decreased from 29% to 18%) [231].

Given the huge task of replacing fossil fuels, contributions are surely required from energy efficiency, renewable energies, and nuclear power, with the mix depending on local preferences. Renewable energy and nuclear power have been limited in part by technical challenges. Nuclear power faces persistent concerns about safety, nuclear waste, and potential weapons proliferation, despite past contributions to mortality prevention and climate change mitigation [232]. Most renewable energies tap diffuse intermittent sources often at a distance from the user population, thus requiring large-scale energy storage and transport. Developing technologies can ameliorate these issues, as discussed below. However, apparent cost is the constraint that prevents nuclear and renewable energies from fully supplanting fossil fuel electricity generation.

Transition to a post-fossil fuel world of clean energies will not occur as long as fossil fuels appear to the investor and consumer to be the cheapest energy. Fossil fuels are cheap only because they do not pay their costs to society and receive large direct and indirect subsidies [233]. Air and water pollution from fossil fuel extraction and use have high costs in human health, food production, and natural ecosystems, killing more than 1,000,000 people per year and affecting the health of billions of people [232], [234], with costs borne by the public. Costs of climate change and ocean acidification, already substantial and expected to grow considerably [26], [235], also are borne by the public, especially by young people and future generations.

Thus the essential underlying policy, albeit not sufficient, is for emissions of CO_2 to come with a price that allows these costs to be internalized within the economics of energy use. Because so much energy is used through expensive capital stock, the price should rise in a predictable way to enable people and businesses to efficiently adjust lifestyles and investments to minimize costs. Reasons for preference of a carbon fee or tax over cap-and-trade include the former's simplicity and relative ease of becoming global [236]. A near-global carbon tax might be achieved, e.g., via a bi-lateral agreement between China and the United States, the greatest emitters, with a border duty imposed on products from nations without a carbon tax, which would provide a strong incentive for other nations to impose an equivalent carbon tax. The suggestion of a carbon fee collected from fossil fuel companies with all revenues distributed to the public on a per capita basis [237] has received at least limited support [238].

Economic analyses indicate that a carbon price fully incorporating environmental and climate damage would be high [239]. The cost of climate change is uncertain to a factor of 10 or more and could be as high as ~$1000/tCO$_2$ [235], [240]. While the imposition of such a high price on carbon emissions is outside the realm of short-term political feasibility, a price of that magnitude is not required to engender a large change in emissions trajectory.

An economic analysis indicates that a tax beginning at $15/tCO$_2$ and rising $10/tCO$_2$ each year would reduce emissions in the U.S. by 30% within 10 years [241]. Such a reduction is more than 10 times as great as the carbon content of tar sands oil carried by the proposed Keystone XL pipeline (830,000 barrels/day) [242]. Reduced oil demand would be nearly six times the pipeline capacity [241], thus the carbon fee is far more effective than the proposed pipeline.

A rising carbon fee is the sine qua non for fossil fuel phase out, but not enough by itself. Investment is needed in RD&D (research, development and demonstration) to help renewable energies and nuclear power overcome obstacles limiting their contributions. Intermittency of solar and wind power can be alleviated with advances in energy storage, low-loss smart electric grids, and electrical vehicles interacting with the grid. Most of today's nuclear power plants have half-century-old technology with light-water reactors [243] utilizing less than 1% of the energy in the nuclear fuel and leaving unused fuel as long-lived nuclear "waste" requiring sequestration for millennia. Modern light-water reactors can employ convective cooling to eliminate the need for external cooling in the event of an anomaly such as an earthquake. However, the long-term future of nuclear power will employ "fast" reactors, which utilize ~99% of the nuclear fuel and can "burn" nuclear waste and excess weapons material [243]. It should be possible to reduce the cost of nuclear power via modular standard reactor design, but governments need to provide a regulatory environment that supports timely construction of approved designs. RD&D on carbon capture and storage (CCS) technology is needed, especially given our conclusion that the current atmospheric CO2 level is already in the dangerous zone, but continuing issues with CCS technology [7], [244] make it inappropriate to construct fossil fuel power plants with a promise of future retrofit for carbon capture. Governments should support energy planning for housing and transportation, energy and carbon efficiency requirements for buildings, vehicles and other manufactured products, and climate mitigation and adaptation in undeveloped countries.

Economic efficiency would be improved by a rising carbon fee. Energy efficiency and alternative low-carbon and no-carbon energies should be allowed to compete on an equal footing, without subsidies, and the public and business community should be made aware that the fee will continually rise. The fee for unconventional fossil fuels, such as oil from tar sands and gas from hydrofracking, should include carbon released in mining and refining processes, e.g., methane leakage in hydrofracking [245]–[249]. If the carbon fee rises continually and predictably, the resulting energy transformations should generate many jobs, a welcome benefit for nations still suffering from long-standing economic recession. Economic modeling shows that about 60% of the public, especially low-income people, would receive more money via a per capita 100% dispersal of the collected fee than they would pay because of increased prices [241].

11.7.3 FAIRNESS: INTERGENERATIONAL JUSTICE AND HUMAN RIGHTS

Relevant fundamentals of climate science are clear. The physical climate system has great inertia, which is due especially to the thermal inertia of the ocean, the time required for ice sheets to respond to global warming, and the longevity of fossil fuel CO_2 in the surface carbon reservoirs (atmosphere, ocean, and biosphere). This inertia implies that there is additional climate change "in the pipeline" even without further change of atmospheric composition. Climate system inertia also means that, if large-scale climate change is allowed to occur, it will be exceedingly long-lived, lasting for many centuries.

One implication is the likelihood of intergenerational effects, with young people and future generations inheriting a situation in which grave consequences are assured, practically out of their control, but not of their doing. The possibility of such intergenerational injustice is not remote – it is at our doorstep now. We have a planetary climate crisis that requires urgent change to our energy and carbon pathway to avoid dangerous consequences for young people and other life on Earth.

Yet governments and industry are rushing into expanded use of fossil fuels, including unconventional fossil fuels such as tar sands, tar shale, shale gas extracted by hydrofracking, and methane hydrates. How can this course be unfolding despite knowledge of climate consequences and evidence that a rising carbon price would be economically efficient and reduce demand for fossil fuels? A case has been made that the absence of effective governmental leadership is related to the effect of special interests on policy, as well as to public relations efforts by organizations that profit from the public's addiction to fossil fuels [237], [250].

The judicial branch of governments may be less subject to pressures from special financial interests than the executive and legislative branches, and the courts are expected to protect the rights of all people, including the less powerful. The concept that the atmosphere is a public trust [251], that today's adults must deliver to their children and future generations an atmosphere as beneficial as the one they received, is the basis for a lawsuit [252] in which it is argued that the U.S. government is obligated to protect the atmosphere from harmful greenhouse gases.

Independent of this specific lawsuit, we suggest that intergenerational justice in this matter derives from fundamental rights of equality and justice. The Universal Declaration of Human Rights [253] declares "All are equal before the law and are entitled without any discrimination to equal protection of the law." Further, to consider a specific example, the United States Constitution provides all citizens "equal protection of the laws" and states that no person can be deprived of "life, liberty or property without due process of law". These fundamental rights are a basis for young people to expect fairness and justice in a matter as essential as the condition of the planet they will inhabit. We do not prescribe the legal arguments by which these rights can be achieved, but we maintain that failure of governments to effectively address climate change infringes on fundamental rights of young people.

Ultimately, however, human-made climate change is more a matter of morality than a legal issue. Broad public support is probably needed to achieve the changes needed to phase out fossil fuel emissions. As with the issue of slavery and civil rights, public recognition of the moral dimensions of human-made climate change may be needed to stir the public's conscience to the point of action.

A scenario is conceivable in which growing evidence of climate change and recognition of implications for young people lead to massive public support for action. Influential industry leaders, aware of the moral issue, may join the campaign to phase out emissions, with more business leaders becoming supportive as they recognize the merits of a rising price on carbon. Given the relative ease with which a flat carbon price can be made international [236], a rapid global emissions phasedown is feasible. As fossil fuels are made to pay their costs to society, energy efficiency and clean energies may reach tipping points and begin to be rapidly adopted.

Our analysis shows that a set of actions exists with a good chance of averting "dangerous" climate change, if the actions begin now. However, we also show that time is running out. Unless a human "tipping point" is reached soon, with implementation of effective policy actions, large irreversible climate changes will become unavoidable. Our parent's generation did not know that their energy use would harm future generations and other life on the planet. If we do not change our course, we can only pretend that we did not know.

11.8 DISCUSSION

We conclude that an appropriate target is to keep global temperature within or close to the temperature range in the Holocene, the interglacial period in which civilization developed. With warming of 0.8°C in the past century, Earth is just emerging from that range, implying that we need to restore the planet's energy balance and curb further warming. A limit of approximately 500 GtC on cumulative fossil fuel emissions, accompanied by a net storage of 100 GtC in the biosphere and soil, could keep global temperature close to the Holocene range, assuming that the net future forcing change from other factors is small. The longevity of global warming (Fig. 9) and the implausibility of removing the warming if it is once allowed to penetrate the deep ocean emphasize the urgency of slowing emissions so as to stay close to the 500 GtC target.

Fossil fuel emissions of 1000 GtC, sometimes associated with a 2°C global warming target, would be expected to cause large climate change with disastrous consequences. The eventual warming from 1000 GtC fossil fuel emissions likely would reach well over 2°C, for several reasons. With such emissions and temperature tendency, other trace greenhouse gases including methane and nitrous oxide would be expected to increase, adding to the effect of CO_2. The global warming and shifting climate zones would make it less likely that a substantial increase in forest and soil carbon could be achieved. Paleoclimate data indicate that slow feedbacks would substantially amplify the 2°C global warming. It is clear that pushing global climate far outside the Holocene range is inherently dangerous and foolhardy.

The fifth IPCC assessment Summary for Policymakers [14] concludes that to achieve a 50% chance of keeping global warming below 2°C equivalent CO_2 emissions should not exceed 1210 GtC, and after accounting for non-CO_2 climate forcings this limit on CO_2 emissions becomes 840 GtC. The existing drafts of the fifth IPCC assessment are not yet approved for comparison and citation, but the IPCC assessment is consistent with studies of Meinshausen et al. [254] and Allen et al. [13], hereafter M2009 and A2009, with which we can make comparisons. We will also compare our conclusions with those of McKibben [255]. M2009 and A2009 appear together in the same journal with the two lead authors on each paper being co-authors on the other paper. McKibben [255], published in a popular magazine, uses quantitative results of M2009 to conclude that most remaining fossil fuel reserves must be left in the ground, if global warming this century is to be kept below 2°C. McKibben [255] has been very successful in drawing public attention to the urgency of rapidly phasing down fossil fuel emissions.

M2009 use a simplified carbon cycle and climate model to make a large ensemble of simulations in which principal uncertainties in the carbon cycle, radiative forcings, and climate response are allowed to vary, thus yielding a probability distribution for global warming as a function of time throughout the 21st century. M2009 use this distribution to infer a limit on total (fossil fuel+net land use) carbon emissions in the period 2000–2049 if global warming in the 21st century is to be kept below 2°C at some specified probability. For example, they conclude that the limit on total 2000–2049 carbon emissions is 1440 $GtCO_2$ (393 GtC) to achieve a 50% chance that 21st century global warming will not exceed 2°C.

A2009 also use a large ensemble of model runs, varying uncertain parameters, and conclude that total (fossil fuel+net land use) carbon emissions of 1000 GtC would most likely yield a peak CO_2-induced warming of 2°C, with 90% confidence that the peak warming would be in the range 1.3–3.9°C. They note that their results are consistent with those of M2009, as the A2009 scenarios that yield 2°C warming have 400–500 GtC emissions during 2000–2049; M2009 find 393 GtC emissions for 2°C warming, but M2009 included a net warming effect of non-CO_2 forcings, while A2009 neglected non-CO_2 forcings.

McKibben [255] uses results of M2009 to infer allowable fossil fuel emissions up to 2050 if there is to be an 80% chance that maximum warming in the 21st century will not exceed 2°C above the pre-industrial level. M2009 conclude that staying under this 2°C limit with 80% probability requires that 2000–2049 emissions must be limited to 656 $GtCO_2$ (179 GtC) for 2007–2049. McKibben [255] used this M2009 result to determine a remaining carbon budget (at a time not specified exactly) of 565 $GtCO_2$ (154 GtC) if warming is to stay under 2°C. Let us update this analysis to the present: fossil fuel emissions in 2007–2012 were 51 GtC [5], so, assuming no net emissions from land use in these few years, the M2009 study implies that the remaining budget at the beginning of 2013 was 128 GtC.

Thus, coincidentally, the McKibben [255] approach via M2009 yields almost exactly the same remaining carbon budget (128 GtC) as our analysis (130 GtC). However, our budget is that required to limit warming to about 1°C (there is a temporary maximum during this century at about 1.1–1.2°C, Fig. 9), while McKibben [255] is allowing global warming to reach 2°C, which we have concluded would be a disaster scenario! This apparently vast difference arises from three major factors.

First, we assumed that reforestation and improved agricultural and forestry practices can suck up the net land use carbon of the past. We estimate net land use emissions as 100 GtC, while M2009 have land use emissions almost twice that large (~180 GtC). We argue elsewhere (see section 14 in Supporting Information of [54]) that the commonly employed net land use estimates [256] are about a factor of two larger than the net land use carbon that is most consistent with observed CO_2 history. However, we need not resolve that long-standing controversy here. The point is that, to make the M2009 study equivalent to ours, negative land use emissions must be included in the 21st century equal to earlier positive land use emissions.

Second, we have assumed that future net change of non-CO_2 forcings will be zero, while M2009 have included significant non-CO_2 forcings. In recent years non-CO_2 GHGs have provided about 20% of the increase of total GHG climate forcing.

Third, our calculations are for a single fast-feedback equilibrium climate sensitivity, 3°C for doubled CO_2, which we infer from paleoclimate data. M2009 use a range of climate sensitivities to compute a probability distribution function for expected warming, and then McKibben [255] selects the carbon emission limit that keeps 80% of the probability distribution below 2°C.

The third factor is a matter of methodology, but one to be borne in mind. Regarding the first two factors, it may be argued that our scenario is optimistic. That is true, but both goals, extracting 100 GtC from the atmosphere via improved forestry and agricultural practices (with possibly some assistance from CCS technology) and limiting additional net change of non-CO_2 forcings to zero, are feasible and probably much easier than the principal task of limiting additional fossil fuel emissions to 130 GtC.

We noted above that reforestation and improving agricultural and forestry practices that store more carbon in the soil make sense for other reasons. Also that task is made easier by the excess CO_2 in the air today, which causes vegetation to take up CO_2 more efficiently. Indeed, this may be the reason that net land use emissions seem to be less than is often assumed.

As for the non-CO_2 forcings, it is noteworthy that greenhouse gases controlled by the Montreal Protocol are now decreasing, and recent agreement has been achieved to use the Montreal Protocol to phase out production of some additional greenhouse gases even though those gases do not affect the ozone layer. The most important non-CO_2 forcing is methane, whose increases in turn cause tropospheric ozone and stratospheric water vapor to increase. Fossil fuel use is probably the largest source of methane [1], so if fossil fuel use begins to be phased down, there is good basis to anticipate that all three of these greenhouse gases could decrease, because of the approximate 10-year lifetime of methane.

As for fossil fuel CO_2 emissions, considering the large, long-lived fossil fuel infrastructure in place, the science is telling us that policy should be set to reduce emissions as rapidly as possible. The most fundamental implication is the need for an across-the-board rising fee on fossil fuel emissions in order to allow true free market competition from non-fossil energy sources. We note that biospheric storage should not be allowed to offset further fossil fuel emissions. Most fossil fuel carbon will remain in the climate system more than 100,000 years, so it is essential to limit the emission of fossil fuel carbon. It will be necessary to have incentives to restore biospheric carbon, but these must be accompanied by decreased fossil fuel emissions.

A crucial point to note is that the three tasks [limiting fossil fuel CO_2 emissions, limiting (and reversing) land use emissions, limiting (and reversing) growth of non-CO_2 forcings] are interactive and reinforcing. In mathematical terms, the problem is non-linear. As one of these climate forcings increases, it increases the others. The good news is that, as one of them decreases, it tends to decrease the others. In order to bestow upon future generations a planet like the one we received, we need to win on all three counts, and by far the most important is rapid phasedown of fossil fuel emissions.

It is distressing that, despite the clarity and imminence of the danger of continued high fossil fuel emissions, governments continue to allow and even encourage pursuit of ever more fossil fuels. Recognition of this reality and perceptions of what is "politically feasible" may partially account for acceptance of targets for global warming and carbon emissions that are well into the range of "dangerous human-made interference" with climate. Although there is merit in simply chronicling what is happening, there is still opportunity for humanity to exercise free will. Thus our objective is to define what the science indicates is needed, not to assess political feasibility. Further, it is not obvious to us that there are physical or economic limitations that prohibit fossil fuel emission targets far lower than 1000 GtC, even targets closer to 500 GtC. Indeed, we suggest that rapid transition off fossil fuels would have numerous near-term and long-term social benefits, including improved human health and outstanding potential for job creation.

A world summit on climate change will be held at United Nations Headquarters in September 2014 as a preliminary to negotiation of a new climate treaty in Paris in late 2015. If this treaty is analogous to the 1997 Kyoto Protocol [257], based on national targets for emission reductions and cap-and-trade-with-offsets emissions trading mechanisms, climate deterioration and gross intergenerational injustice will be practically guaranteed. The palpable danger that such an approach is conceivable is suggested by examination of proposed climate policies of even the most forward-looking of nations. Norway, which along with the other Scandinavian countries has been among the most ambitious and successful of all nations in reducing its emissions, nevertheless approves expanded oil drilling in the Arctic and development of tar sands as a majority owner of Statoil [258]–[259]. Emissions foreseen by the Energy Perspectives of Statoil [259], if they occur, would approach or exceed 1000 GtC and cause dramatic climate change that would run out of control of future generations. If, in contrast, leading nations agree in 2015 to have internal rising fees on carbon with border duties on products from nations without a carbon fee, a foundation would be established for phaseover to carbon free energies and stable climate.

REFERENCES

1. Intergovernmental Panel on Climate Change (2007) Climate Change 2007: Physical Science Basis, Solomon, S, Dahe, Q, Manning M, Chen Z, Marquis M, et al., editors. Cambridge Univ. Press: New York 2007; 996 pp.
2. Hansen J, Sato M, Ruedy R, Nazarenko L, Lacis A, et al.. (2005) Efficacy of climate forcings. J Geophys Res 110, D18104, doi:10.1029/2005JD005776.

3.   Archer D (2005) Fate of fossil fuel CO2 in geologic time. J Geophys Res 110: C09S05. doi: 10.1029/2004jc002625

4.   BP Statistical Review of World Energy 2012 (http://www.bp.com/sectionbodycopy. do?categoryId=7500&contentId=7068481).

5.   Boden TA, Marland G, Andres RJ (2012) Global, Regional, and National Fossil-Fuel CO2 Emissions. Carbon Dioxide Information Analysis Center, Oak Ridge National Laboratory, U.S. Department of Energy, Oak Ridge, Tenn., U.S.A. doi 10.3334/CDIAC/00001_V2012.

6.   United Nations Framework Convention on Climate Change (1992) Available: http://www.unfccc.int.

7.   Energy Information Administration (EIA) (2011) International Energy Outlook Available: http://www.eia.gov/forecasts/ieo/pdf/0484.Pdf accessed Sept 2011.

8.   German Advisory Council on Global Change (GAC)(2011) World in Transition - A Social Contract for Sustainability. Available: http://www.wbgu.de/en/flagship-reports/fr-2011-a-social-contract/. Accessed Oct 2011.

9.   Global Energy Assessment (GEA) (2012) Toward a Sustainable Future. Johanson TB, Patwardhan E, Nakićenović N, editors. Cambridge: Cambridge University Press.

10. Randalls S (2010) History of the 2°C climate target. WIREs Clim Change 1, 598–605.

11. Copenhagen Accord (2009) United Nations Framework Convention on Climate Change, Draft decision −/CP.15 FCCC/CP/2009/L.7 18 December 2009.

12. Matthews HD, Gillett NP, Stott PA, Zickfeld K (2009) The proportionality of global warming to cumulative carbon emissions. Nature 459: 829–832. doi: 10.1038/nature08047

13. Allen MR, Frame DJ, Huntingford C, Jones CD, Lowe JA, et al.. (2009) Warming caused by cumulative carbon emissions towards the trillionth tonne. Nature 458, 1163–1166.

14. Intergovernmental Panel on Climate Change (2013) Approved Summary for Policymakers of full draft report of Climate Change 2013: Physical Science Basis, Stocker T, Dahe Q, Plattner G-K, coordinating lead authors, available: http://www.ipcc.ch/report/ar5/wg1/#.UlCweRCvHMM.

15. Intergovernmental Panel on Climate Change (2007) Climate Change 2007: Mitigation of Climate Change. Metz B, Davidson OR, Bosch PR, Dave R, Meyer LA, editors. Cambridge: Cambridge University Press.

16. Hansen J, Ruedy R, Sato M, Lo K (2010) Global Surface Temperature Change. Rev Geophys 48: RG4004. doi: 10.1029/2010rg000345

17. Meehl GA, Arblaster JM, Marsh DR (2013) Could a future "Grand Solar Minimum" like the Maunder Minimum stop global warming? Geophys Res Lett 40, 1789–1793.

18. Kosaka Y, Xie SP (2013) Recent global-warming hiatus tied to equatorial Pacific surface cooling. Nature published online 28 August doi:10.1038/nature12534.

19. Intergovernmental Panel on Climate Change (2001) Climate Change 2001: The Scientific Basis. Houghton JT, MacCarthy JJ, Metz M, editors. Cambridge: Cambridge University Press.

20. Schneider SH, Mastrandrea MD (2005) Probabilistic assessment "dangerous" climate change and emissions pathways. Proc Natl Acad Sci USA 102: 15728–15735. doi: 10.1073/pnas.0506356102

21. Stroeve JC, Kattsov V, Barrett A, Serreze M, Pavlova T, et al. (2012) Trends in Arctic sea ice extent from CMIP5, CMIP3 and observations. Geophys Res Lett 39: L16502. doi: 10.1029/2012gl052676

22. Rampal P, Weiss J, Dubois C, Campin JM (2011) IPCC climate models do not capture Arctic sea ice drift acceleration: Consequences in terms of projected sea ice thinning and decline. J Geophys Res 116: C00D07. doi: 10.1029/2011jc007110

23. Shepherd A, Ivins ER, Geruo A, Barletta VR, Bentley MJ, et al. (2012) A reconciled estimate of ice-sheet mass balance. Science 338: 1183–1189. doi: 10.1126/science.1228102

24. Rignot E, Velicogna I, van den Broeke MR, Monaghan A, Lenaerts J (2011) Acceleration of the contribution of the Greenland and Antarctic ice sheets to sea level rise. Geophys Res Lett 38: L05503–L05508. doi: 10.1029/2011gl046583

25. Hanna E, Navarro FJ, Pattyn F, Domingues CM, Fettweis X, et al. (2013) Ice-sheet mass balance and climate change. Nature 498: 51–59. doi: 10.1038/nature12238

26. Intergovernmental Panel on Climate Change (2007) Climate Change 2007: Impacts, Adaptation and Vulnerability. Parry, ML, Canziani O, Palutikof J, van der Linden P, Hanson C, editors. Cambridge: Cambridge University Press.

27. Rabatel A, Francou B, Soruco A, Gomez J, Caceres B, et al. (2013) Current state of glaciers in the tropical Andes: a multi-century perspective on glacier evolution and climate change. The Cryosphere 7: 81–102. doi: 10.5194/tc-7-81-2013

28. Sorg A, Bolch T, Stoffel M, Solomina O, Beniston M (2012) Climate change impacts on glaciers and runoff in Tien Shan (Central Asia). Nature Clim Change 2, 725–731.

29. Yao T, Thompson L, Yang W, Yu W, Gao Y, et al.. (2012) Differrent glacier status with atmospheric circulations in Tibetan Plateau and surroundings. Nature Clim Change 2, 663–667.

30. Barnett TP, Pierce DW, Hidalgo HG, Bonfils C, Santer BD, et al. (2008) Human-induced changes in the hydrology of the western United States. Science 319: 1080–1083. doi: 10.1126/science.1152538

31. Kaser G, Grosshauser M, Marzeion B (2010) Contribution potential of glaciers to water availability in different climate regimes. Proc Natl Acad Sci USA 107: 20223–20227. doi: 10.1073/pnas.1008162107

32. Vergara W, Deeb AM, Valencia AM, Bradley RS, Francou B, et al. (2007) Economic impacts of rapid glacier retreat in the Andes. EOS Trans Amer. Geophys Union 88: 261–268. doi: 10.1029/2007eo250001

33. Held IM, Soden BJ (2006) Robust responses of the hydrological cycle to global warming. J Clim 19: 5686–5699. doi: 10.1175/jcli3990.1

34. Seidel DJ, Fu Q, Randel WJ, Reichler TJ (2008) Widening of the tropical belt in a changing climate. Nat Geosci 1: 21–24.

35. Davis SM, Rosenlof KH (2011) A multi-diagnostic intercomparison of tropical width time series using reanalyses and satellite observations. J Clim doi: 10.1175/JCLI-D-1111–00127.00121.

36. Liu J, Song M, Hu Y, Ren X (2012) Changes in the strength and width of the Hadley circulation since 1871. Clim Past 8: 1169–1175. doi: 10.5194/cp-8-1169-2012

37. Dai A (2013) Increasing drought under global warming in observations and models. Nature Clim Change 3, 52–58.
38. Westerling AL, Hidalgo HG, Cayan DR, Swetnam TW (2006) Warming and earlier spring increase western US forest wildfire activity. Science 313: 940–943. doi: 10.1126/science.1128834
39. Bruno JF, Selig ER (2007) Regional decline of coral cover in the Indo-Pacific: timing, extent, and subregional comparisons. Plos One 2: e711. doi: 10.1371/journal.pone.0000711
40. Hoegh-Guldberg O, Mumby PJ, Hooten AJ, Steneck RS, Greenfield P, et al. (2007) Coral reefs under rapid climate change and ocean acidification. Science 318: 1737–1742. doi: 10.1126/science.1152509
41. Veron JE, Hoegh-Guldberg O, Lenton TM, Lough JM, Obura DO, et al. (2009) The coral reef crisis: The critical importance of <350 ppm CO2. Mar Pollut Bull 58: 1428–1436. doi: 10.1016/j.marpolbul.2009.09.009
42. Parmesan C, Yohe G (2003) A globally coherent fingerprint of climate change impacts across natural systems. Nature 421: 37–42. doi: 10.1038/nature01286
43. Parmesan C (2006) Ecological and evolutionary responses to recent climate change. Ann Rev Ecol Evol S 37: 637–669. doi: 10.1146/annurev.ecolsys.37.091305.110100
44. Poloczanska ES, Brown CJ, Sydeman WJ, Kiessling W, Schoeman DS, et al.. (2013) Global imprint of climate change on marine life. Nature Clim Change doi:10.1038/NCLIMATE1958.
45. Rahmstorf S, Coumou D (2011) Increase of extreme events in a warming world. Proc Natl Acad Sci USA 108: 17905–17909. doi: 10.1073/pnas.1101766108
46. Hansen J, Sato M, Ruedy R (2012) Perception of climate change. Proc Natl Acad Sci USA 109 14726–14727.
47. Lewis SC, Karoly DJ (2013) Anthropogenic contributions to Australia's record summer temperatures of 2013. Geophys Res Lett (in press).
48. Jouzel J, Masson-Delmotte V, Cattani O, Dreyfus G, Falourd S, et al. (2007) Orbital and millennial Antarctic climate variability over the past 800,000 years. Science 317: 793–796. doi: 10.1126/science.1141038
49. Masson-Delmotte V, Stenni B, Pol K, Braconnot P, Cattani O, et al. (2010) EPICA Dome C record of glacial and interglacial intensities. Quat Sci Rev 29: 113–128. doi: 10.1016/j.quascirev.2009.09.030
50. Zachos J, Pagani M, Sloan L, Thomas E, Billups K (2001) Trends, rhythms, and aberrations in global climate 65 Ma to present. Science 292: 686–693. doi: 10.1126/science.1059412
51. Rohling EJ, Sluijs A, Dijkstra HA, Kohler P, van de Wal RSW, et al. (2012) Making sense of palaeoclimate sensitivity. Nature 491: 683–691. doi: 10.1038/nature11574
52. Hansen J, Sato M, Russell G, Kharecha P (2013) Climate sensitivity, sea level, and atmospheric CO2. Philos Trans R Soc A 371: 20120294, 2013.
53. Foster GL, Rohling EJ (2013) Relationship between sea level and climate forcing by CO2 on geological timescales. Proc Natl Acad Sci USA doi:10.1073/pnas.1216073110.
54. Hansen J, Sato M, Kharecha P, Beerling D, Berner R, et al. (2008) Target Atmospheric CO2: Where Should Humanity Aim? The Open Atmospheric Science Journal 2: 217–231. doi: 10.2174/1874282300802010217

55. Marcott SA, Shakun JD, Clark PU, Mix AC (2013) A reconstruction of regional and global temperature for the last 11,300 years. Science 339: 1198–1201. doi: 10.1126/science.1228026

56. Pagani M, Liu ZH, LaRiviere J, Ravelo AC (2010) High Earth-system climate sensitivity determined from Pliocene carbon dioxide concentrations. Nat Geosci 3: 27–30. doi: 10.1038/ngeo724

57. Meyssignac B, Cazenave A (2012) Sea level: a review of present-day and recent-past changes and variability. J Geodynamics 58, 96–109.

58. Berger AL (1978) Long term variations of daily insolation and quaternary climate changes. J Atmos Sci 35: 2362–2367. doi: 10.1175/1520-0469(1978)035<2362:ltv odi>2.0.co;2

59. Hansen J, Sato M, Kharecha P, Russell G, Lea DW, et al. (2007) Climate change and trace gases. Phil Tran Roy Soc 365: 1925–1954. doi: 10.1098/rsta.2007.2052

60. Kohler P, Fischer H, Joos F, Knutti R, Lohmann G, et al. (2010) What caused Earth's temperature variations during the last 800,000 years? Data-based evidence on radiative forcing and constraints on climate sensitivity. Quat Sci Rev 29: 29–145. doi: 10.1016/j.quascirev.2009.09.026

61. Masson-Delmotte V, Stenni B, Pol K, Braconnot P, Cattani O, et al. (2010) EPICA Dome C record of glacial and interglacial intensities. Quat Sci Rev 29: 113–128. doi: 10.1016/j.quascirev.2009.09.030

62. Rohling EJ, Medina-Elizalde M, Shepherd JG, Siddall M, Stanford JD (2011) Sea surface and high-latitude temperature sensitivity to radiative forcing of climate over several glacial cycles. J Clim doi: 10.1175/2011JCLI4078.1171.

63. Beerling DJ, Royer DL (2011) Convergent Cenozoic CO_2 history. Nat Geosci 4: 418–420. doi: 10.1038/ngeo1186

64. Hansen J, Sato M, Kharecha P, Schuckmann K (2011) Earth's Energy Imbalance and Implications. Atmos Chem Phys 11: 1–29. doi: 10.5194/acp-11-13421-2011

65. Levitus S, Antonov JI, Wang J, Delworth TL, Dixon KW, Broccoli AJ (2001) Anthropogenic warming of earth's climate system. Science 292, 267–270.

66. Roemmich D, Gilson J (2009) The 2004–2008 mean and annual cycle of temperature, salinity, and steric height in the global ocean from the Argo Program. Prog Oceanogr 82: 81–100. doi: 10.1016/j.pocean.2009.03.004

67. Lyman JM, Good SA, Gouretski VV, Ishii M, Johnson GC, et al. (2010) Robust warming of the global upper ocean. Nature 465: 334–337. doi: 10.1038/nature09043

68. Barker PM, Dunn JR, Domingues CM, Wijffels SE (2011) Pressure Sensor Drifts in Argo and Their Impacts. J Atmos Ocean Tech 28: 1036–1049. doi: 10.1175/2011jte-cho831.1

69. Levitus S, Antonov JI, Boyer TP, Baranova OK, Garcia HE, et al.. (2012) World ocean heat content and thermosteric sea level change (0–2000 m), 1955–2010, Geophys Res Lett 39, L10603.

70. von Schuckmann K, LeTraon P-Y (2011) How well can we derive Global Ocean Indicators from Argo data? Ocean Sci 7: 783–391. doi: 10.5194/os-7-783-2011

71. Frohlich C (2006) Solar irradiance variability since 1978. Space Sci Rev 125: 53–65. doi: 10.1007/s11214-006-9046-5

72. Murphy DM, Solomon S, Portmann RW, Rosenlof KH, Forster PM, Wong T (2009) An observationally based energy balance for the Earth since 1950. J Geophys Res 114: D17107. doi: 10.1029/2009jd012105

73. Mishchenko MI, Cairns B, Kopp G, Schueler CF, Fafaul BA, et al. (2007) Accurate monitoring of terrestrial aerosols and total solar irradiance: introducing the Glory mission. B Am Meteorol Soc 88: 677–691. doi: 10.1175/bams-88-5-677

74. Economist (2013) Beijing's air pollution: blackest day. Economist, 14 January 2013. Available at: http://www.economist.com/blogs/analects/2013/01/beijings-air-pollution.

75. Hansen J, Sato M, Ruedy R, Lacis A, Oinas V (2000) Global warming in the twenty-first century: An alternative scenario. Proc Natl Acad Sci USA 97: 9875–9880. doi: 10.1073/pnas.170278997

76. Bond T, Doherty SJ, Fahey DW, Forster PM, Berntsen T, et al.. (2013) Bounding the role of black carbon in the climate system: a scientific assessment. J Geophys Res (in press).

77. Smith JB, Schneider SH, Oppenheimer M, Yohe GW, Hare W, et al.. (2009) Assessing dangerous climate change thorough an update of the Intergovernmental Panel on Climate Change (IPCC) "reasons of concern". Proc Natl Acad Sci USA 106, 4133–4137.

78. Hearty PJ, Hollin JT, Neumann AC, O'Leary MJ, McCulloch M (2007) Global sea-level fluctuations during the Last Interglaciation (MIS 5e). Quaternary Sci Rev 26: 2090–2112. doi: 10.1016/j.quascirev.2007.06.019

79. Kopp RE, Simons FJ, Mitrovica JX, Maloof AC, Oppenheimer M (2009) Probabilistic assessment of sea level during the last interglacial stage. Nature 462: 863–867. doi: 10.1038/nature08686

80. Dutton A, Lambeck K (2012) Ice volume and sea level during the last interglacial. Science 337: 216–219. doi: 10.1126/science.1205749

81. Rohling EJ, Grant K, Hemleben C, Siddall M, Hoogakker BAA, et al. (2008) High rates of sea-level rise during the last interglacial period. Nat Geosci 1: 38–42. doi: 10.1038/ngeo.2007.28

82. Thompson WG, Curran HA, Wilson MA, White B (2011) Sea-level oscillations during the last interglacial highstand recorded by Bahamas corals. Nat Geosci 4: 684–687. doi: 10.1038/ngeo1253

83. Blanchon P, Eisenhauer A, Fietzke J, Volker L (2009) Rapid sea-level rise and reef back-stepping at the close of the last interglacial highstand. Nature 458: 881–884. doi: 10.1038/nature07933

84. Hearty PJ, Neumann AC (2001) Rapid sea level and climate change at the close of the Last Interglaciation (MIS 5e): evidence from the Bahama Islands. Quaternary Sci Rev 20: 1881–1895. doi: 10.1016/s0277-3791(01)00021-x

85. O'Leary MJ, Hearty PJ, Thompson WG, Raymo ME, Mitrovica X, et al.. (2013) Ice sheet collapse following a prolonged period of stable sea level during the Last Interglacial. Nature Geosci., published online 28 July. doi:10.1038/NGEO1890.

86. Raymo ME, Mitrovica JX, O'Leary MJ, DeConto RM, Hearty P (2011) Departures from eustasy in Pliocene sea-level records. Nat Geosci 4: 328–332. doi: 10.1038/ngeo1118

87. Naish TR, Wilson G (2009) Constraints on the amplitude of Mid-Pliocene (3.6–2.4 Ma) eustatic sea-level fluctuations from the New Zealand shallow-marine sediment record. Philos Trans R Soc A 367: 169–187. doi: 10.1098/rsta.2008.0223

88. Hill DJ, Haywood DM, Hindmarsh RCM, Valdes PJ (2007) Characterizing ice sheets during the Pliocene: evidence from data and models. In: Williams M, Haywood AM, Gregory J, Schmidt DN, editors. Deep-Time Perspectives on Climate Change: Marrying the Signal from Computer Models and Biological Proxies. London: Micropalaeont Soc Geol Soc. 517–538.

89. Dwyer GS, Chandler MA (2009) Mid-Pliocene sea level and continental ice volume based on coupled benthic Mg/Ca palaeotemperatures and oxygen isotopes. Phil Trans R Soc A 367: 157–168. doi: 10.1098/rsta.2008.0222

90. Rignot E, Bamber JL, van den Broeke MR, Davis C, Li Y, et al. (2008) Recent Antarctic ice mass loss from radar interferometry and regional climate modelling. Nat Geosci 1: 106–110. doi: 10.1038/ngeo102

91. NEEM community members (2013) Eemian interglacial reconstructed from a Greenland folded ice core. Nature 493: 489–494.

92. Hughes T (1972) Is the West Antarctic ice sheet disintegrating? ISCAP Bulletin, no. 1, Ohio State Univ.

93. Oppenheimer M (1999) Global warming and the stability of the West Antarctic ice sheet. Nature 393: 325–332.

94. Bentley CR (1997) Rapid sea-level rise soon from West Antarctic ice sheet collapse? Science 275: 1077–1078. doi: 10.1126/science.275.5303.1077

95. Vermeer M, Rahmstorf S (2009) Global sea level linked to global temperature. Proc Nat Acad Sci USA 106: 21527–21532. doi: 10.1073/pnas.0907765106

96. Grinsted A, Moore J, Jevrejeva S (2010) Reconstructing sea level from paleo and projected temperatures 200 to 2100 AD. Clim Dyn 34: 461–472. doi: 10.1007/s00382-008-0507-2

97. Hansen JE (2005) A slippery slope: How much global warming constitutes "dangerous anthropogenic interference"? Clim Chg 68: 269–279. doi: 10.1007/s10584-005-4135-0

98. Hansen J (2007) Scientific reticence and sea level rise. Env Res Lett 2: 024002. doi: 10.1088/1748-9326/2/2/024002

99. Tedesco M, Fettweis X, Mote T, Wahr J, Alexander P, et al.. (2012) Evidence and analysis of 2012 Greenland records from spaceborne observations, a regional climate model and reanalysis data. Cryospre Discuss 6, 4939–4976.

100. Levi BG (2008) Trends in the hydrology of the western US bear the imprint of man-made climate change. Physics Today 61: 16–18. doi: 10.1063/1.2911164

101. Hansen J, Sato M, Ruedy R, Lo K, Lea DW, et al. (2006) Global temperature change. Proc Natl Acad Sci USA 103: 14288–14293. doi: 10.1073/pnas.0606291103

102. Burrows MT, Schoeman DS, Buckley LB, Moore P, Poloczanska ES, et al. (2011) The Pace of Shifting Climate in Marine and Terrestrial Ecosystems. Science 334: 652–655. doi: 10.1126/science.1210288

103. Hoegh-Guldberg O, Bruno JF (2010) The Impact of Climate Change on the World's Marine Ecosystems. Science 328: 1523–1528. doi: 10.1126/science.1189930

104. Seimon TA, Seimon A, Daszak P, Halloy SRP, Sdchloegel LM, et al. (2007) Upward range extension of Andean anurans and chytridiomycosis to extreme eleva-

tions in response to tropical deglaciation. Global Change Biol 13: 288–299. doi: 10.1111/j.1365-2486.2006.01278.x

105. Pounds JA, Fogden MPL, Campbell JH (1999) Biological response to climate change on a tropical mountain. Nature 398: 611–615. doi: 10.1038/19297

106. Pounds JA, Bustamante MR, Coloma LA, Consuegra JA, Fogden MPL, et al. (2006) Widespread amphibian extinctions from epidemic disease driven by global warming. Nature 439: 161–167. doi: 10.1038/nature04246

107. Alford RA, Bradfield KS, Richards SJ (2007) Ecology: Global warming and amphibian losses. Nature 447: E3–E4. doi: 10.1038/nature05940

108. Rosa ID, Simoncelli F, Fagotti A, Pascolini R (2007) Ecology: The proximate cause of frog declines? Nature 447: E4–E5. doi: 10.1038/nature05941

109. Pounds JA, Bustamante MR, Coloma LA, Consuegra JA, Fogden MPL, et al. (2007) Ecology – Pounds et al reply. Nature 447: E5–E6. doi: 10.1038/nature05942

110. Mahlstein I, Daniel JS, Solomon S (2013) Pace of shifts in climate regions increases with global temperature. Nature Clim Change doi:10.1038/nclimate1876.

111. Olson S, Hearty P (2010) Predation as the primary selective force in recurrent evolution of gigantism in Poecilozonites land snails in Quaternary Bermuda. Biol Lett 6, 807–810.

112. Hearty P, Olson S (2010) Geochronology, biostratigraphy, and chaning shell morphology in the land snail subgenus Poecilozonites during the Quaternary of Bermuda. Palaeog Plaeocl Palaeoeco 293, 9–29.

113. Olson S, Hearty P (2003) Probably extirpation of a middle Pleistocene breeding colony of Short-tailed Albatross (Phoebastria albatrus) on Bermuda by a +20 m interglacial sea-level rise. Proc Natl Acad Sci USA 100, 12825–12829.

114. Taylor J, Braithwaite C, Peake J, Arnold E (1979) Terrestrial fauna and habitats of Aldabra during the lat Pleistocene. Phil Trans Roy Soc Lon B 286, 47–66.

115. 2010 IUCN Red List of Threatened Species (http://www.iucnredlist.org/details/9010/0).

116. Butchart SHM, Walpole M, Collen B, van Strein A, Scharlemann JPW, et al. (2010) Global biodiversity: indicators of recent declines. Science 328: 1164–1168. doi: 10.1126/science.1187512

117. Raup DM, Sepkoski JJ (1982) Mass Extinctions in the Marine Fossil Record. Science 215: 1501–1503. doi: 10.1126/science.215.4539.1501

118. Barnosky AD, Matzke N, Tomiya S, Wogan GOU, Swartz B, et al. (2011) Has the Earth's sixth mass extinction already arrived? Nature 471: 51–57. doi: 10.1038/nature09678

119. Reaka-Kudla ML (1997) Global biodiversity of coral reefs: a comparison with rainforests. In: Reaka-Kudla ML, Wilson DE, Wilson EO, editors. Biodiversity II: Understanding and Protecting Our Biological Resources. Washington, DC: Joseph Henry Press. 83–108.

120. Caldeira K, Wickett ME (2003) Oceanography: Anthropogenic carbon and ocean pH. Nature 425: 365–365. doi: 10.1038/425365a

121. Raven J, Caldeira K, Elderfield H, Hoegh-Guldberg O, Liss P, et al.. (2005) Ocean acidification due to increasing atmospheric carbon dioxide. London: Royal Society.

122. Pelejero C, Calvo E, Hoegh-Guldberg O (2010) Paleo-perspectives on ocean acidification. Trends Ecol Evol 25: 332–344. doi: 10.1016/j.tree.2010.02.002

123. Hoegh-Guldberg O (1999) Climate change, coral bleaching and the future of the world's coral reefs. Mar Freshwater Res 50: 839–866. doi: 10.1071/mf99078

124. Death G, Lough JM, Fabricius KE (2009) Declining Coral Calcification on the Great Barrier Reef. Science 323: 116–119. doi: 10.1126/science.1165283

125. Seager R, Naik N, Vogel L (2012) Does global warming cause intensified interannual hydroclimate variability? J Clim 25: 3355–3372. doi: 10.1175/jcli-d-11-00363.1

126. Held IM, Delworth TL, Lu J, Findell KL, Knutson TR (2005) Simulation of Sahel drought in the 20th and 21st centuries. Proc Natl Acad Sci USA 102: 17891–17896. doi: 10.1073/pnas.0509057102

127. Groisman PY, Knight RW, Easterling DR, Karl TR, Hegerl GC, et al. (2005) Trends in intense precipitation in the climate record. J Clim 18: 1326–1350. doi: 10.1175/jcli3339.1

128. Alexander LV, Zhang X, Peterson TC, Caesar J, Gleason B, et al. (2006) Global observed changes in daily climate extremes of temperature and precipitation. J Geophys Res 111: D05109. doi: 10.1029/2005jd006290

129. Min SK, Zhang X, Zwiers FW, Hegerl GC (2011) Human contribution to more-intense precipitation extremes. Nature 470: 378–381. doi: 10.1038/nature09763

130. Dai A (2011) Drought under global warming: a review. WIREs Clim Change 2: 45–65. doi: 10.1002/wcc.81

131. Briffa KR, van der Schrier G, Jones PD (2009) Wet and dry summers in Europe since 1750: evidence of increasing drought. Int J Climatol 29: 1894–1905. doi: 10.1002/joc.1836

132. Sheffield J, Wood EF, Roderick ML (2012) Little change in global drought over the past 60 years. Nature 491: 435–438. doi: 10.1038/nature11575

133. Robine JM, Cheung SL, Le Roy S, Van Oyen H, Griffiths C, et al. (2008) Death toll exceeded 70,000 in Europe during the summer of 2003. Cr Biol 331: 171–175. doi: 10.1016/j.crvi.2007.12.001

134. Barriopedro D, Fischer EM, Luterbacher J, Trigo R, Garcia-Herrera R (2011) The Hot Summer of : Redrawing the Temperature Record Map of Europe. Science 332: 220–224. doi: 10.1126/science.1201224

135. Stott PA, Stone DA, Allen MR (2004) Human contribution to the European heatwave of 2003. Nature 432: 610–614. doi: 10.1038/nature03089

136. Fritze JG, Blashki GA, Burke S, Wiseman J (2008) Hope, despair and transformation: climate change and the promotion of mental health and well-being. International J Mantal Health Sys 7: 2–13. doi: 10.1186/1752-4458-2-13

137. Searle K, Gow K (2010) Do concerns about climate change lead to distress? International J Clim Change Strat Manag 2: 362–378. doi: 10.1108/17568691011089891

138. Hicks D, Bord A (2001) Learning about global issues: why most educators only make things worse. Envir Education Res 7: 413–425. doi: 10.1080/13504620120081287

139. Gottlieb D, Bronstein P (1996) Parent's perceptions of children's worries in a changing world. J Genetic Psychol 157: 104–118. doi: 10.1080/00221325.1996.9914849

140. Chen Y, Ebenstein A, Greenstone M, Li H (2013) Evidence on the impact of sustained exposure to air pollution on life expectancy from China's Huai River policy. Proc Natl Acad Sci USA www.pnas.org/cgi/doi/10.1073/pnas.1300018110.

141. Davidson DJ, Andrews J (2013) Not all about consumption. Science 339, 1286–1287.

142. Murphy DJ, Hall CAS (2011) Energy return on investment, peak oil, and the end of economic growth. Ann New York Acad Sci 1219, 52–72.

143. Palmer MA, Bernhardt ES, Schlesinger WH, Eshleman KN, Foufoula-Georgiou E, et al.. (2010) Mountaintop mining consequences. Science 327, 148–149.

144. Allan JD (2004) Landscapes and riverscapes: The influence of land use on stream ecosystems. Annu Rev Eco Evol Syst 35, 257–284.

145. McCormick BC, Eshleman KN, Griffith JL, Townsend PA (2009) Detection of flooding responses at the river basin scale enhanced by land use change. Water Resoures Res 45, W08401.

146. Pond GJ, Passmore ME, Borsuk FA, Reynolds L, Rose CJ (2008) Downstream effects of mountaintop coal mining: comparing biological conditions using family- and genus-level macroinvertebrate bioassessment tools. J N Am Benthol Soc 27, 717–737.

147. McAuley SD, Kozar MD (2006) Ground-water quality in unmined areas and near reclaimed surface coal mines in the northern and central Appalachian coal regions, Pennsylvania and West Virginia, http://pubs.usgs.gov/sir/2006/5059/pdf/sir2006-5059.pdf.

148. Negley TL, Eshleman KN (2006) Comparison of stormflow responses of surface-mined and forested watersheds in the Appalachian Mountains, USA. Hydro Process 20, 3467–3483.

149. Simmons JA, Currie WS, Eshleman KN, Kuers K, Monteleone S, et al.. (2008) Forest to reclaimed mine use change leads to altered ecosystem structure and function. Ecolog Appl 18, 104–118.

150. Energy Resources and Conservation Board (2012) Alberta's energy reserves 2011 and supply/demand outlook – Appendix D, www.ercb.ca/sts/ST98/ST98-2012.pdf.

151. Jordaan SM, Keith DW, Stelfox B (2009) Quantifying land use of oil sands production: a life cycle perspective. Environ Res Lett 4, 1–15.

152. Yeh S, Jordaan SM, Brandt AR, Turetsky MR, Spatari S, et al.. (2010) Land use greenhouse gas emissions from conventional oil production and oil sands. Environ Sci Technol 44, 8766–8772.

153. Charpentier AD, Bergerson JA, MacLean HL (2009) Understanding the Canadian oil snads industry's greenhouse gas emissions. Environ Res Lett 4, 014005, 11 pp.

154. Johnson EA, Miyanishi K (2008) Creating new landscapes and ecosystems: the Alberta oil sands. Ann NY Acad Sci 1134, 120–145.

155. Allen L, Cohen MJ, Abelson D, Miller B (2011) Fossil fuels and water quality, in The World's Water, Springer, 73–96.

156. Rooney RC, Bayley SE, Schindler DW (2011) Oil sands mining and reclamation cause massive loss of peatland and stored carbon. Proc Natl Acad Sci USA, www.pnas.org/cgi/doi/10.1073/pnas.1117693108.

157. Kurek J, Kirk JL, Muir DCG, Wang X, Evans MS, et al.. (2013) Legacy of a half century of Athabasca oil sands development recorded by lake ecosystems. Proc Natl Acad Sci USA www.pnas.org/cgi/doi/10.1073/pnas.1217675110.

158. Kelly EN, Schindler DW, Hodson PV, Short JW, Radmanovich R, et al.. (2010) Oil sands development contributes elements toxic at low concentrations to the Athabasca River and its tributaries. Proc Natl Acad Sci USA 107, 16178–16183.

159. Schmidt CW (2011) Blind Rush? Shale gas boom proceeds amid human health questions. Environ Health Perspec 119, A348–A353.
160. Kargbo DM, Wilhelm RG, Caampbell DJ (2010) Natural gas plays in the Marcellus shale: challenges and potential opportunities. Environ Sci Technol 44, 5679–5684.
161. Gregory KB, Vidic RD, Dzombak DA (2011) Water management challenges associated with the production of shale gas by hydraulic fracturing. Elements 7, 181–186.
162. Riverkeeper (2011) Fractured communities: case studies of the environmental impacts of industrial gas drilling. http://tinyurl.com/373rpp4.
163. Osborn SG, Vengosh A, Warner NR, Jackson RB (2011) Methane contamination of drinking water accompanying gas-well drilling and hydraulic fracturing. Proc Natl Acad Sci USA 108, 8172–8176.
164. O'Sullivan F, Paltsev S (2012) Shale gas production: potential versus actual greenhouse gas emissions. Environ Res Lett 7, 044030.
165. Allen L, Cohen MJ, Abelson D, Miller B (2011) Fossil fuels and water quality, in The World's Water, Springer, New York, 73–96.
166. Hansen J, Sato M, Ruedy R, Lacis A, Asamoah K, et al.. (1997) Forcings and chaos in interannual to decadal climate change. J Geophys Res 102, 25679–25720.
167. Hansen J, Sato M (2004) Greenhouse gas growth rates. Proc Natl Acad Sci USA 101: 16109–16114. doi: 10.1073/pnas.0406982101
168. Archer D (2007) Methane hydrate stability and anthropogenic climate change. Biogeosciences 4: 521–544. doi: 10.5194/bg-4-521-2007
169. Joos F, Bruno M, Fink R, Siegenthaler U, Stocker TF, et al. (1996) An efficient and accurate representation of complex oceanic and biospheric models of anthropogenic carbon uptake. Tellus B Chem Phys Meterol 48: 397–417. doi: 10.3402/tellusb.v48i3.15921
170. Kharecha PA, Hansen JE (2008) Implications of "peak oil" for atmospheric CO2 and climate. Global Biogeochem Cy 22: GB3012. doi: 10.1029/2007gb003142
171. Stocker TF (2013) The closing door of climate targets. Science 339, 280–282.
172. Stocker BD, Strassmann K, Joos F (2011) Sensitivity of Holocene atmospheric CO2 and the modern carbon budget to early human land use: analyses with a process-based model. Biogeosciences 8: 69–88. doi: 10.5194/bg-8-69-2011
173. Sarmiento JL, Gloor M, Gruber N, Beaulieu C, Jacobson AR, et al.. (2010) Trends and regional distributions of land and ocean carbon sinks. Biogeosci 7, 2351–2367.
174. Hillel D, Rosenzweig C, editors (2011) Handbook of Climate Change and Agroecosystems: Impacts, Adaptation and Mitigation. London: Imperial College Press.
175. Lamb D (2011) Regreening the Bare Hills. New York: Springer. 547 p.
176. Smith P (2012) Agricultural greenhouse gas mitigation potential globally, in Europe and in the UK: what have we learned in the last 20 years? Global Change Biol 18: 35–43. doi: 10.1111/j.1365-2486.2011.02517.x
177. Rockstrom J, Falkenmark M, Karlberg L, Hoff H, Rost S, Gerten D (2009) Future water availaility for global food production: The potential of greenwater for increasing resilience to global change. Water Resour Res 45, W00A12, doi:10.1029/2007WR006767.
178. Smith P, Gregory PJ, van Vuuren D, Obersteiner M, Havlik P, et al. (2010) Competition for land. Philos T R Soc B 365: 2941–2957. doi: 10.1098/rstb.2010.0127

179. Malhi Y (2010) The carbon balance of tropical forest regions, 1990–2005. Curr Op Environ Sustain 2, 237–244.

180. Bala G, Caldeira K, Wickett M, Phillips TJ, Lobell DB, et al. (2007) Combined climate and carbon-cycle effects of large-scale deforestation. Proc Natl Acad Sci USA 104: 6550–6555. doi: 10.1073/pnas.0608998104

181. Bonan GB (2008) Forests and climate change: Forcings, feedbacks, and the climate benefits of forests. Science 320: 1444–1449. doi: 10.1126/science.1155121

182. Zomer RJ, Trabucco A, Bossio DA, Verchot LV (2008) Climate change mitigation: A spatial analysis of global land suitability for clean development mechanism afforestation and reforestation. Agriculture Ecosystems & Environment 126: 67–80. doi: 10.1016/j.agee.2008.01.014

183. Tilman D, Hill J, Lehman C (2006) Carbon-negative biofuels from low-input high-diversity grassland biomass. Science 314: 1598–1600. doi: 10.1126/science.1133306

184. Fargione J, Hill J, Tilman D, Polasky S, Hawthorne P (2008) Land clearing and the biofuel carbon debt. Science 319: 1235–1238. doi: 10.1126/science.1152747

185. Searchinger T, Heimlich R, Houghton RA, Dong F, Elobeid A, et al. (2008) Use of US croplands for biofuels increases greenhouse gases through emissions from land-use change. Science 319: 1238–1240. doi: 10.1126/science.1151861

186. Stehfest E, Bouwman L, van Vuuren DP, den Elzen MGJ, Eickhout B, et al. (2009) Climate benefits of changing diet. Clim Chg 95: 83–102. doi: 10.1007/s10584-008-9534-6

187. Hansen J, Kharecha P, Sato M (2013) Climate forcing growth rates: doubling down on our Faustian bargain. Envir Res Lett 8,011006, doi:10.1088/1748–9326/8/1/011006.

188. Earth System Research Laboratory (2013) www.esrl.noaa.gov/gmd/ccgg/trends/.

189. Frohlich C, Lean J (1998) The Sun's total irradiance: cycles and trends in the past two decades and associated climate change uncertainties. Geophys Res Lett 25, 4377–4380.

190. Hansen J, Sato M, Ruedy R, Kharecha P, Lacis A, et al. (2007) Dangerous human-made interference with climate: a GISS modelE study. Atmos Chem Phys 7, 2287–2312.

191. http://www.columbia.edu/~mhs119/Solar/and original sources given therein.

192. Eddy JA (1776) The Maunder Minimum. Science 192, 1189–1202.

193. Lean J, Beer J, Bradley R (1995) Reconstruction of solar irradiance since 1610: implications for climate change. Geophys Res Lett 22: 3195–3198. doi: 10.1029/95gl03093

194. Jones GS, Lockwood M, Stott PA (2012) What influence will future solar activity changes over the 21st century have on projected global near-surface temperature changes? J Geophys Res 117: D05103. doi: 10.1029/2011jd017013

195. Lu Z, Zhang Q, Streets DG (2011) Sulfur dioxide and primary carbonaceous aerosol emissions in China and India, 1996–2010. Atmos Chem Phys 11: 9839–9864. doi: 10.5194/acp-11-9839-2011

196. Robock A (2000) Volcanic eruptions and climate. Rev Geophys 38: 191–219. doi: 10.1029/1998rg000054

197. Gleckler PJ, Wigley TML, Santer BD, Gregory JM, AchutaRao K, et al. (2006) Krakatoa's signature persists in the ocean. Nature 439: 675. doi: 10.1038/439675a

198. Solomon S, Daniel JS, Sanford TJ, Murphy DM, Plattner GK, et al. (2010) Persistence of climate changes due to a range of greenhouse gases. Proc Natl Acad Sci USA 107: 18354–18359. doi: 10.1073/pnas.1006282107

199. Broecker WS, Bond G, Klas M, Bonani G, Wolfi W (1990) A salt oscillator in the glacial North Atlantic? Paleoeanography 5, 469–477.

200. Hansen JE, Sato M (2012) Paleoclimate implications for human-made climate change, in Climate Change: Inferences from Paleoclimate and Regional Aspects. A. Berger, F. Mesinger, and D. Šijački, Eds. Springer, 21–48, doi:10.1007/978-3-7091-0973-1-2.

201. Eby M, Zickfeld K, Montenegro A, Archer D, Meissner KJ, et al.. (2009) Lifetime of anthropogenic climate change: millennial time-scales of potential CO2 and surface temperature perturbations. J Clim 22, 2501–2511.

202. DeAngelis H, Skvarca P (2003) Glacier surge after ice shelf collapse. Science 299, 1560–1562.

203. Pritchard HD, Ligtenberg SRM, Fricker HA, Vaughan DG, van den Broeke, et al.. (2012) Antarctic ice-sheet loss driven by basal melting of ice shelves. Nature 484, 502–505.

204. Broecker WS, Peng TH (1982) Tracers in the Sea, Eldigio, Palisades, New York, 1982.

205. Kennett JP, Stott LD (1991) Abrupt deep-sea warming, paleoceanographic changes and benthic extinctions at the end of the Paleocene. Nature 353, 225–229.

206. Ridgwell A (2007) Interpreting transient carbonate compensation depth changes by marine sediment core modeling. Paleoceanography 22, PA4102.

207. Zeebe RE, Zachos JC, Dickens GR (2009) Carbon dioxide forcing alone insufficient to explain Palaeocene-Eocene Thermal Maximum warming. Nature Geosci 2, 576–580.

208. Berner RA (2004) The Phanerozoic Carbon Cycle: CO2 and O2, Oxford Univ. Press, New York.

209. Max MD (2003) Natural Gas Hydrate in Oceanic and Permafrost Environments. Boston: Kluwer Academic Publishers.

210. Kvenvolden KA (1993) Gas Hydrates - Geological Perspective and Global Change. Rev Geophys 31: 173–187. doi: 10.1029/93rg00268

211. Dickens GR, O'Neil JR, Rea DK, Owen RM (1995) Dissociation of oceanic methane hydrate as a cause of the carbon isotope excursion at the end of the Paleocene. Paleoceanography 10, 965–971.

212. DeConto RM, Galeotti S, Pagani M, Tracy D, Schaefer K, et al.. (2012) Past extreme wrming events linked to massive carbon release from thawing permafrost. Nature 484, 87–91.

213. Walter K, Zimov S, Chanton J, Verbyla D, Chapin F (2006) Methane bubbling from Siberian thaw lakes as a positive feedback to climate warming. Nature 443: 71–75. doi: 10.1038/nature05040

214. Shakhova N, Semiletov I, Salyuk A, Yusupov V, Kosmac D, et al. (2010) Extensive Methane Venting to the Atmosphere from Sediments of the East Siberian Arctic Shelf. Science 327: 1246–1250. doi: 10.1126/science.1182221

215. O'Connor FM, Boucher O, Gedney N, Jones CD, Folberth GA, et al.. (2010) Possible role of wetlands, permafrost, and methane hydrates in the methane cycle under future climate change: a review. Rev Geophys 48, RG4005.

216. Lunt DJ, Haywood AM, Schmidt GA, Salzmann U, Valdes PJ, et al. (2010) Earth system sensitivity inferred from Pliocene modelling and data. Nature Geosci 3: 60–64. doi: 10.1038/ngeo706

217. Harris NL, Brown S, Hagen SC, Saatchi SS, Petrova S, et al.. (2012) Baseline map of carbon emissions from deforestation in tropical regions. Science 336, 1573–1576.

218. Matthews HD, Keith DW (2007) Carbon-cycle feedbacks increase the likelihood of a warmer future. Geophys Res Lett 34: L09702. doi: 10.1029/2006gl028685

219. Friedlingstein P, Cox P, Betts R, Bopp L, von Bloh W, et al.. (2006) Climate-Carbon Cycle feedback analysis: results fromC4MIP model intercomparison. J Clim 19, 3337–3353.

220. Huntingford C, Zelazowski P, Galbraith D, Mercado LM, Sitch S, et al.. (2013) Simulated resilience of tropical rainforests to CO2-induced climate change. Nature Geosciece, doi:10.1038/ngeo1741.

221. Naik V, Mauzerall D, Horowitz L, Schwarzkopf MD, Ramaswamy V, et al.. (2005) Net radiative forcing due to changes in regional emissions of tropospheric ozone precursors. J Geophys Res 110, D24, doi:10.1029/2005JD005908.

222. Beerling DJ, Stevenson DS, Valdes PJ (2011) Enhanced chemistry-climate feedbacks in past greenhouse worlds. Proc Natl Acad Sci USA 108, 9770–9775.

223. Shepherd J (2009) Geoengineering the climate: science, governance and uncertainty. London: The Royal Society, London, 84 pp. available http://www.royalsociety.org.

224. Budyko MI (1977) Climate changes. American Geophysical Union, Washington, DC, p. 244.

225. Robock A (2008) 20 reasons why geoengineering may be a bad idea. Bull Atom Sci 64, 14–18.

226. Keith DW, Ha-Duong M, Stolaroff JK (2006) Climate strategy with CO2 capture from the air. Clim. Chg 74: 17–45. doi: 10.1007/s10584-005-9026-x

227. House KZ, Baclig AC, Ranjan M, van Nierop EA, Wilcox J, et al.. (2011) Economic and energetic analysis of capturing CO2 from ambient air. Proc Natl Acad Sci USA 108, 20428–20433.

228. APS (2011) Direct Air Capture of CO2 with Chemicals: A Technology Assessment for the APS Panel on Public Affairs. American Physical Society. Available: http://www.aps.org/policy/reports/assessments/upload/dac2011.pdf. Accessed Jan 11, 2012.

229. U.S. Energy Information Administration (2012) Annual Energy Review 2011, 370 pp., www.eia.gov/aer.

230. Pauwelyn J (2012) Carbon leakage measures and border tax adjustments under WTO law, in Research Handbook on Environment, Health and the WTO 48–49, eds. Provost Cand Van Calster G.

231. International Energy Agency (2012), "World energy balances", IEA World Energy Statistics and Balances (database). doi: 10.1787/data-00512-en. Accessed Mar. 2013.

232. Kharecha P, Hansen J (2013) Prevented mortality and greenhouse gas emissions from historical and projected nuclear power. Envir Sci Tech 47: 4889–4895. doi: 10.1021/es3051197

233. International Energy Agency (2012), World Energy Outlook 2012. 690pp. OECD/ IEA (http://www.worldenergyoutlook.org/publications/weo-2012/).

234. Cohen AJ, Ross Anderson H, Ostro B, Pandey KD, Krzyzanowski M, et al. (2005) The Global Burden of Disease Due to Outdoor Air Pollution. J Toxicol Environ Health A 68: 1301–1307. doi: 10.1080/15287390590936166

235. Ackerman F, Stanton EA (2012) Climate Risks and Carbon Prices: Revising the Social Cost of Carbon. Economics E-journal 6, 2012–10.5018/economics-ejournal. ja.2012–10.

236. Hsu S-L (2011) The Case for a Carbon Tax. Washington, DC: Island Press.

237. Hansen J (2009) Storms of My Grandchildren. New York: Bloomsbury. 304 pp.

238. Lochhead C (2013) George Shultz pushes for carbon tax. San Francisco Chronicle, 8 March.

239. Stern N (2007) Stern Review on the Economics of Climate Change Cambridge: Cambridge University Press.

240. Ackerman F, DeCanio S, Howarth R, Sheeran K (2009) Limitations of integrated assessment models of climate change. Clim Change 95: 297–315. doi: 10.1007/ s10584-009-9570-x

241. Komanoff C (2011) 5-Sector Carbon Tax Model: http://www.komanoff.net/fossil/ CTC_Carbon_Tax_Model.xls. Accessed December 25, 2011.

242. United States Department of State (2011) Final Environmental Impact Statement. Available: http://www.state.gov/r/pa/prs/ps/2011/08/171084.htm. Accessed 09 February 2013.

243. Till CE, Chang YI (2011) Plentiful energy: the story of the integral fast reactor United States: Charles E. Till and Yoon Il Chang. 116 pp.

244. Kramer D (2012) Scientists poke holes in carbon dioxide sequestration. Phys Today 65: 22–24. doi: 10.1063/pt.3.1672

245. Tollefson J (2012) Air sampling reveals high emissions from gas fields. Nature 482, 139–140.

246. Tollefson J (2013) Methane leaks erode green credentials of natural gas. Nature 493, 12.

247. Petron G, Frost GJ, Miller BR, Hirsch AL, Montzka SA, et al.. (2012) Hydrocarbon emissions characterizations in the Colorado Front Range. J Geophys Res 117, D04304.

248. Petron G, Frost GJ, Trainer MK, Miller BR, Dlugokencky EJ, et al.. (2013) Reply to comment on "Hydrocarbon emissions characterization in the Colirado Front Range – A pilot study" by Michael A. Levi. J Geophys Res 118, D018487.

249. Alvarez RA, Pacala SW, Winebrake JJ, Chameides WL, Hamburg SP (2012) Greater focus needed on methane leakage from natural gas infrastructure. Proc Natl Acad Sci USA.

250. Oreskes N, Conway EM (2010) Merchants of Doubt: How a Handful of Scientists Obscured the Truth on Issues from Tobacco Smoke to Global Warming. New York: Bloomsbury Press. 355 pp. merchantsofdoubt.org.

251. Wood MC (2009) Atmospheric Trust Litigation. In: Burns WCG, Osofsky HM, editors. Adjudicating Climate Change: Sub-National, National, And Supra-National Approaches. Cambridge: Cambridge University Press. 99–125. Available: http://www.law.uoregon.edu/faculty/mwood/docs/atmospheric.pdf.

252. Alec L v. Jackson DDC, No. 11-CV-02235, 12/14/11 (United States District Court, District of Columbia).

253. Universal Declaration of Human Rights (http://www.un.org/en/documents/udhr/).

254. Meinshausen M, Meinshausen N, Hare W, Raper SCB, Frieler K, et al.. (2009) Greenhouse gas emission targets for limiting global warming to 2°C. Nature 458, 1158–1162.

255. McKibben B (2012) Global warming's terrifying new math. Rolling Stone, August 2.

256. Houghton RA (2003) Revised estimates of the annual net flux of carbon to the atmosphere from changes in land use and land management 1850–2000. Tellus B 55: 378–390. doi: 10.1034/j.1600-0889.2003.01450.x

257. http://unfccc.int/kyoto_protocol/items/2830.php.

258. http://www.regjeringen.no/en/dep/md/documents-and-publications/government-propositions-and-reports-/reports-to-the-storting-white-papers-2/2011-2012/report-no-21-2011-2012.html?id=707321.

259. http://www.statoil.com/en/NewsAndMedia/News/EnergyPerspectives/Pages/default.asp

There are several supplemental files that are not available in this version of the article. To view this additional information, please use the citation on the first page of this chapter.

Author Notes

CHAPTER 4

Authors Information
HR is a professor at Indian Institute of Technology. PCJ is a scientist at Central Institute of Agricultural Engineering. SSJ is a systems engineer at John Deere India Pvt. Ltd.

Acknowledgments
To carry out this research work, the financial help provided by Petroleum Conservation Research Association, Government of India, New Delhi is sincerely acknowledged.

Competing Interests
The authors declare that they have no competing interests.

Author Contributions
All the authors - HR, PCJ, and SSJ - collectively carried out experiments, collected data, and analyzed them. HR wrote the manuscript. All authors read and approved the final manuscript.

CHAPTER 5

Acknowledgments
This study was supported by mechanics of agricultural machinery department of Razi university of Iran. The authors would like to acknowledge contributions of Razi university of Iran and mechanics of agricultural machinery laboratory.

CHAPTER 6

Acknowledgments

The author wishes to thank Mr. Siva Thanapal for the help in preparation of this editorial. A part of the analysis was supported with a DOT Sun grant program administered through Oklahoma State University, OK, USA.

CHAPTER 8

Acknowledgments

Support by the Academy of Sciences for the Developing World (TWAS), Polytechnic of Namibia, International Centre for Environmental and Nuclear Sciences (ICENS) and Faculty of Science and Technology, University of the West Indies (Mona Campus) is highly acknowledged.

CHAPTER 9

Acknowledgments

We thank Terry Chapin, Chris Costello, Jon Foley, David Hughes, Amy Mall, Roxanne Marino, Nathan Phillips, and Drew Shindell for comments and advice. Funding was provided by the Park Foundation and endowments given by David. R. Atkinson and Dwight C. Baum to Cornell University. The authors have no financial conflict of interests.

CHAPTER 10

Acknowledgments

We gratefully acknowledge Dr. Stephen Moore for his valuable comments on the manuscript and the Kyoto City Environmental Policy Bureau for their generous support of this research.

CHAPTER 11

Funding

Funding came from: NASA Climate Research Funding, Gifts to Columbia University from H.F. ("Gerry") Lenfest, private philanthropist (no web site, but seehttp://en.wikipedia.org/wiki/H._F._Lenfest), Jim Miller, Lee Wasserman (Rockefeller Family Fund) (http://www.rffund.org/), Flora Family Foundation (http://www.florafamily.org/), Jeremy Grantham, ClimateWorks and the Energy Foundation provided support for Hansen's Climate Science, Awareness and Solutions program at Columbia University to complete this research and publication. The funders had no role in study design, data collection and analysis, decision to publish, or preparation of the manuscript.

Competing Interests

The authors have declared that no competing interests exist.

Acknowledgments

We greatly appreciate the assistance of editor Juan A. Añel in achieving requisite form and clarity for publication. The paper is dedicated to Paul Epstein, a fervent defender of the health of humans and the environment, who graciously provided important inputs to this paper while battling late stages of non-Hodgkin's lymphoma. We thank David Archer, Inez Fung, Charles Komanoff and two anonymous referees for perceptive helpful reviews and Mark Chandler, Bishop Dansby, Ian Dunlop, Dian Gaffen Seidel, Edward Greisch, Fred Hendrick, Tim Mock, Ana Prados, Stefan Rahmstorf, Rob Socolow and George Stanford for helpful suggestions on a draft of the paper.

Author Contributions

Conceived and designed the experiments: JH PK MS. Performed the experiments: MS PK. Wrote the paper: JH. Wrote the first draft: JH. All authors made numerous critiques and suggested specific wording and references: JH PK MS VM-D FA DJB PJH OHG SLH CP JR EJR JS PS KS LVS KvS JCZ. Especially: PK MS VM-D.

Index

T - #0828 - 101024 - C320 - 229/152/14 - PB - 9781774636787 - Gloss Lamination